Technology in Our Time

Volume I

Doing Digital Media

Revised Edition

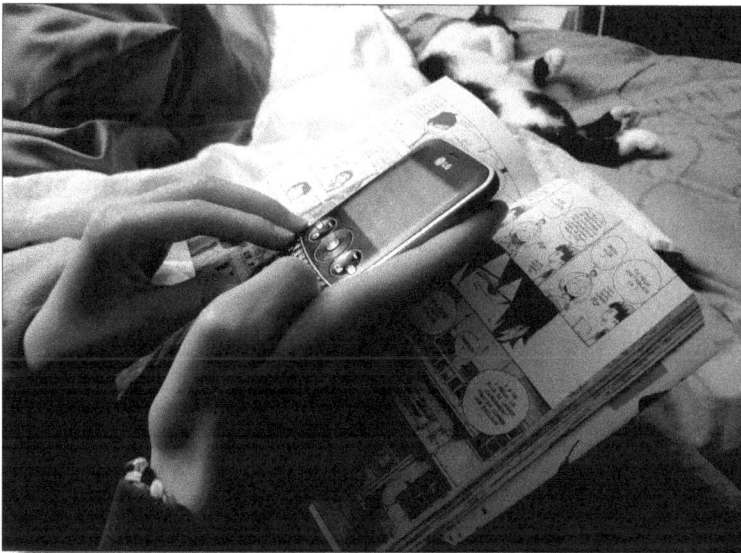

Edited by Laura Robinson
Santa Clara University

Associate Editor	Media Editor
Jeremy Schulz	Myles James Anthony Sansone
UC Berkeley	*Santa Clara University*

cognella™
academic publishing

Bassim Hamadeh, CEO and Publisher
Michael Simpson, Vice President of Acquisitions
Jamie Giganti, Managing Editor
Jess Busch, Graphic Design Supervisor
Jessica Knott, Project Editor
Luiz Ferreira, Licensing Associate

First published in the United States of America in 2013 by Cognella, Inc.

title page/back cover image: © Olybrius (CC BY 3.0) at http://commons.wikimedia.org/wiki/File:Teenage_girl_texting_while_reading_a_manga_1.jpg.

Printed in the United States of America

ISBN: 978-1-62661-065-1 (pbk)/ 978-1-62661-278-5 (br)

cognella™
academic publishing
www.cognella.com   800-200-3908

Contents

Doing Digital Media

Introductory Ideas

By Laura Robinson

*D**oing Digital Media* opens a window onto a broad panorama of topics central to the study of new media. This book invites readers to consider how digital technologies suffuse social environments in a variety of ways. The fundamental means through which we form relationships as individuals, in groups, and across communities are increasingly mediated by digital technologies. The selections examine the impact of new media on our identities, homes, relationships, mobility, social problems, and privacy. Across these topics, the readings explore how various digital media affect the channels in which we interact and communicate, the kinds of roles we adopt, and the technologies we use to we conduct ourselves as members of communities and groups. Each thematic section presents insights into the manifold ways that digital media facilitate, shape, and constrain our everyday lives.

The chapters raise profound and important questions about the positive and negative consequences of living in a social world dominated by new forms of communication. While the direct and indirect influences of new media differ for each of us depending on a variety of factors such as personal viewpoints and tastes, as well as access and usage practices, most of us cannot afford to ignore the pervasive presence of new media. Equipped with our iPhones and smartphones, as well as our iPads and laptops, we are increasingly wired and plugged into the grid. As technology grows more pervasive in our lives, it is important that we leave space and time for reflection on the forces that are acting upon us, as well as the choices we have the power to make as individuals.

Quotes and Questions

Section 1: *Who We Say We Are*

Our readings begin with the section *Who We Say We Are*. In *The Presentation of Self in the Age of Social Media* Hogan writes, "Presentation of self is becoming increasingly popular as a means for explaining differences in meaning and activity of online participation." Hogan introduces the concepts of exhibitioner and curator in order to explain self-presentation online: "Many online sites today set up situations where people can continually update information associated with their profiles. The capabilities of exhibition sites allow a person to be found when others want to look rather than when the person is able to be present and perform."

Ask yourself:
Define and apply Hogan's concepts of exhibitioner and curator to your own digital experiences. Quoting the text, analyze key points in the chapter including symbolic interactionism, identity, and performance as they relate to Goffman. Critically consider the following and make explicit linkages to the reading: How do you weigh information and exchanges on social network sites (SNS) such as Facebook compared to face-to-face (F2F)? What efforts do you make to craft your SNS identity? Do your SNS profiles mirror your offline self? What percentage of your Facebook friends do you communicate with on a daily or weekly basis? On SNS, do people judge each other based on number of contacts or "friends"? Is this valid? Has the definition of "friend" changed because of social media?

The next reading by Marwick and Boyd examines identity work in *I Tweet Honestly, I Tweet Passionately: Twitter Users, Context Collapse, and the Imagined Audience*. While Twitter creates a space for personal interaction, at the same time tweeting may generate oversharing and privacy concerns. Marwick and Boyd argue that although social media technologies bring us together, these technologies also necessitate different interactional and self-presentation techniques than are used in face-to-face conversation: "In combining public-facing and interpersonal interaction, the networked audience creates new opportunities for connection, as well as new tensions and conflicts."

Ask yourself:
Define and apply Marwick and Boyd's concepts of context collapse and micro-celebrity to your own digital experiences. Quoting the text, analyze key points in the chapter including the imagined audience as it relates to context. Critically consider the following and make explicit linkages to the reading: Do you think politicians or public figures have an obligation to use Twitter to communicate with the public? What costs and benefits does Twitter provide for ordinary individuals compared to celebrities? How can oversharing cause issues for anyone who tweets unwisely? Should tweets be protected by the first amendment? Do you tweet? If not, why not? If so, what compels you to tweet?

Section 2: *Where We Live*

The following section, *Where We Live*, opens with *The Digital Home: A New Locus of Social Science Research*. Holohan, Chin, Calaghan, and Mühlau write: "The Internet has revolutionized how we as a society access information and … it has changed the nature of our homes by becoming integrated in appliances and on our televisions, thus creating the idea of a 'Digital Home.'" According to the authors, digital homes empower occupants by allowing them to "customize and control the functionality of their homes to meet their individual needs." With Pervasive Interactive Programming (PiP), digital homes allow for user-inspired innovation in their iSpaces such that individuals may customize their homes according to their personal tastes.

Ask yourself:
Define and apply Holohan et al.'s concepts of customization and functionality to your own digital experiences. Quoting the text, analyze key points in the chapter including user-inspired innovation and digital lifestyles as they relate to the iSpace. Critically consider the following and make explicit linkages to the reading: What would you like to digitize in your living space, personal space, or car? How would digitization enhance your use of these spaces? What are the potential negative consequences of digital homes? How might digitizing the home open the door to monitoring or privacy invasion? Would you want a digital home if all of your activities were used as "data" for third parties?

We move from the house to the neighborhood in Hampton's *Neighborhoods in the Network Society: The e-Neighbors Study*. Hampton's study explores the potential for digital technologies to enhance communication in living situations. While suburban neighborhoods were the most likely to use the digital media gifted by the study, the study indicates that digital media have the power to act as an antidote to social isolation. Comparing highly wired and less-wired neighborhoods, the study "supports the hypothesis that Internet use is increasingly embedded into neighborhood networks."

Ask yourself:
Define and apply Hampton's concepts of e-neighbors and networked neighbors to your own digital experiences. Quoting the text, analyze key points in the chapter including closeness and in-person contact as they relate to the networked society. Critically consider the following and make explicit linkages to the reading: In your home neighborhood how often do neighbors communicate? In person or via mediated communication? For what reasons? Do you think that digital technologies are making people less likely to engage in face-to-face interactions with their neighbors? Is this true for you? Has the internet changed the way you communicate with neighbors or other people? In your opinion, do digital technologies enhance or detract from neighborly communication? Why or why not?

Section 3: *Love and...*

In the next section, *Love and...*, Pascoe's chapter *Intimacy* explores how digital media impacts relationships among teens. Her work illuminates how teens use digital media for normative and nonnormative intimacy practices. She writes that "youth use three primary technologies—mobile phones (though many do still use home phones), instant messaging (IM), and social network ties." Pascoe finds that these technologies make it harder for parents to keep tabs on their children's relationships. At the same time, they play important roles in teen coupling rituals.

Ask yourself:
Define and apply Pascoe's concepts of intimacy practices and courtship rituals to your own digital experiences. Quoting the text, analyze key points in the chapter including meeting, flirting, going out, and breaking up as they relate to teens' normative media practices. Critically consider the following and make explicit linkages to the reading: What compels a couple to become "Facebook official"? If it isn't FB official, is the relationship real? How does the use of social media change during different relationship phases: flirting, courting, committing, breaking up, etc.? What positive or other negative behaviors are facilitated by the use of social media in relationships?

In the next reading, DeMasi explores the burgeoning role of digital venues in *Shopping for Love: Online Dating and the Making of a Cyber Culture of Romance*. According to DeMasi: "The tremendous expansion of online personals, along with the public pronouncements of the people who use them, suggest that technologically mediated dating is now a socially acceptable method for finding intimate partners." Today, online dating no longer carries stigma and shame. Rather, it has become a type of information-seeking practice allowing potential partners to "shop" for one another: "Online dating transforms the search for intimate partners into a consumer activity."

Ask yourself:
Define and apply DeMasi's concept of the commercialization of intimacy to your own digital experiences. Quoting the text, analyze key points in the chapter including normalization and efficiencies as they relate to a cyberculture of romance. Critically consider the following and make explicit linkages to the reading: Do you think online dating facilitates long-term relationships, flings, or both? Would you ever use online dating? Under what circumstances? Why or why not? Is online dating a "consumer activity" in the sense that individuals can "shop" around?

Section 4: *Mobility*

Section 4, *Mobility*, is launched by Hjorth's chapter *Domesticating New Media: A Discussion on Locating Mobile Media*. The 24/7 availability of mobile media may be described as "the twenty-first century's equivalent to the Swiss army knife" in that "mobile media encompasses multiple forms of media including camera, gaming platform, MP3 player, and Internet portal." While ubiquitous connectivity has many positive consequences, there are negative consequences as well: "Work becomes mobile; labor is on a perpetual drip. We are supposed to be available at all times, perpetually connected." In this sense, the liberating promises of mobile technologies can also bind us in unexpected ways.

Ask yourself:
Define and apply Hjorth's concepts of domesticating and locating to your own digital experiences. Quoting the text, analyze key points in the chapter including banality and glaze spaces as they relate to mobility. Critically consider the following and make explicit linkages to the reading: How many hours a day are you wired? Via how many devices? Does being perpetually connected liberate or constrain individuals? Who do you want to be able to contact you at any time: family, friends, co-workers, employers, etc.? How many minutes, hours, or days can you be unwired before you feel the itch to use mobile media? Have you ever gone without digital media, especially SNS, for more than a day? What happened? Did you experience any benefits from your media fasting?

Valcanis contributes the chapter: *An iPhone in Every Hand: Media Ecology, Communication Structures, and the Global Village*. He claims that "a computer or like device is now a near universal fixture in Western homes, much like the television and the telephone before it" such that "sight and sound are global in extent." He presents a vision in which digital technologies surmount barriers of time and space to connect much of humankind.

Ask yourself:
Define and apply Valcanis' concepts of media ecology and communication structures to your own digital experiences. Quoting the text, analyze key points in the chapter including space and time as they relate to the global village. Critically consider the following and make explicit linkages to the reading: In what ways has the computer changed our social practices? Do you think local, national, or transnational cultures transform the use of IT? Or do you think that IT erases cultural differences? Both? How does your culture influence your day-to-day use of mobile devices? Have you ever lived in another culture? If so how was the use of IT different or the same?

The section closes with Rich Ling's *Mobile Telephony and Mediated Ritual Interaction*. Ling takes a critical approach to mobile telephony as a form of ritual interaction. His work exposes to view the dual nature of mobile connectivity. While the "mobile phone has extended the range of group interaction," at the same time it has created "various situations in which the use of a mobile phone seemed to tear at the fiber of society."

Ask yourself:
Define and apply Ling's concepts of remote interaction and co-present interaction to your own digital experiences. Quoting the text, analyze key points in the chapter including negotiation and mobile telephony as they relate to mediated ritual interaction. Critically consider the following and make explicit linkages to the reading: What implicit "rules" govern your cell phone use? Do you think that the ability to text has increased or decreased group interaction? Do you prefer to text or to talk on your phone? Why? Has texting changed the way you employ other forms of communication? What has increased? Decreased? What situations call for texting, talking, and/or media consumption?

Section 5: *The Dark Side*

In the subsequent section, *The Dark Side*, we encounter the digitization of two social problems that also exist offline: bullying and miscommunications. In *Online Communication and Negative Social Ties*, Mesch and Talmud seek to "sort out what is new about cyberbullying from what is not and belongs to more traditional forms of peer harassment known before the advent of information society." Their work identifies the possibility for digital media to augment the potential negative effects of social networks, especially on youth: "Cyberbullying is less likely to be a new behavior than aggression which has acquired an additional medium—cyberspace."

Ask yourself:
Define and apply Mesch and Talmud's concept of negative social ties to your own digital experiences. Quoting the text, analyze key points in the chapter including the social network perspective and youth networks as they relate to cyberbullying. Critically consider the following and make explicit linkages to the reading: Is bullying worse or merely different online or offline? Do SNS create fertile ground for cyberbullying? What can individuals, groups, and political bodies do to prevent cyberbullying? Does cyberbullying change the way people present themselves online? Does using digital media allow for more or less protection from bullies?

In *Putting Social Context into Text: The Semiotics of E-mail Exchange*, Menchik and Tian reveal the unintended miscommunications that can occur over email. They offer several explanations. One possibility is that email "is thought to allow less warmth and to employ fewer senses and cues. "Another possibility calls on communication theorists who "emphasize e-mail's negative effect on interaction quality and assign responsibility to the fact that the communication partner is absent." In either case, they make important observations about emotional cuing and how digital communications can go awry.

Ask yourself:
Define and apply Menchik and Tian's concepts of absence of topical interest and associated argument to your own digital experiences. Quoting the text, analyze key points in the chapter including interest, commitment, and incentives as they relate to email communication. Critically consider the following and make explicit linkages to the reading: Do you agree that email allows for reduced emotional cuing? What miscommunications have occurred in your use of email, texting, or use of SNS? Of the various digital communication media you use, which do you believe mitigates such misunderstandings?

Section 6: *TMI*

Rainie and Wellman's chapter *TMI* begins the section *Too Much Information*. They indicate how new media offers us both more information and more information channels. Yet, with this increasingly mobile connectivity "there is too much information to monitor and digest." As a consequence: "The unprecedented abundance of information that permeates the networked individual's life can often be difficult and stressful to manage." According to Rainie and Wellman: "To deal with TMI, networked individuals employ a number of strategies that range in complexity to cope with and manage the information overload." Some of these strategies include using tags, search engines, and bookmarks. They also make several important points including: "Networked individuals are aware of the costs that come with giving unfettered access to their personal information online and thus adjust their online behavior accordingly."

Ask yourself:
Define and apply Rainie and Wellman's concept of the networked individual to your own digital experiences. Quoting the text, analyze key points in the chapter including veillance of personal information, surveillance, coveillance, and sousveillance as they relate to TMI. Critically consider the following and make explicit linkages to the reading: Do you feel that your digital devices provide access to too much information? Is it difficult to manage many sources of information simultaneously? What strategies do you employ to assess and regulate your information intake? Do you have different strategies for different digital life realms: friends, family, work, entertainment, etc.?

Cell Phones and Email by Nippert-Eng presents how individuals navigate privacy. She argues that: "The number and kinds of demands for attention that we are likely to receive at any given time, in any given place, are much greater when these technologies are in use than when they are not." In addition, Nippert-Eng points to the additional demands produced by mobility: "Mobile communication technologies not only facilitate this burgeoning request for attention, they add another twist. They let others outside our immediate physical grasp reach us." Her chapter makes significant contributions such as: "People associated with one aspect of our lives are now suddenly requesting our attention with increasing regularity in all kinds of places and times other than those in which they traditionally might have been expected to appear."

Ask yourself:
Define and apply Nippert-Eng's concepts of selective concealment and disclosure to your own digital experiences. Quoting the text, analyze key points in the chapter including assigning priority and filtering as they relate to privacy. Critically consider the following and make explicit linkages to the reading: How do you protect your privacy on various digital media? Do you use different strategies for different people and/or groups? How aware do you think most people are about monitoring their online privacy using privacy settings on Facebook, Twitter, Instagram, etc.? How are social media erasing our assumptions about privacy?

The section closes with Lincoln's *FYI: TMI: Toward a holistic social theory of information overload*. Lincoln writes: "Both the popular and academic presses warn of the growing problems created by an ever–increasing flow of information. The Economist (2010) warns us of 'monstrous amounts of data'..." To grapple with TMI, he makes several critiques of existing explanations of TMI that examine the interplay of social dynamics or rely on economic models of rationality. Rather he advocates a holistic understanding of information overload with attention to social context.

Ask yourself:
Define and apply Lincoln's concepts of information overload and information person to your own digital experiences. Quoting the text, analyze key points in the chapter including the history, symptoms, and effects of information overload as they relate to the proposed holistic social theory of this phenomenon. Critically consider the following and make explicit linkages to the reading: How is our society permeated by information overload? What are the consequences? What do you do to limit the vast amount of information confronting you? Is multi-tasking an effective way to deal with information overload or does it create additional problems?

Section 1

Who We Say We Are

The Presentation of Self in the Age of Social Media

Distinguishing Performances and Exhibitions Online

By Bernie Hogan

> All the world's a stage, And all the men and women merely players: They have their exits and their entrances; And one man in his time plays many parts.
> —Shakespeare, *As You Like It*, 2/7

There is a distinct irony in Shakespeare's claim, as spoken by Jacques in *As You Like It*. Shakespeare is not remembered for his charisma, his looks, or his wit at parties but for his voluminous plays and sonnets. The world is not only a stage but also a library and a gallery. We do not merely move through life's stages, as Jacques's monologue suggests, but leave a multitude of data traces as we go. In an era of social media, these data traces do not merely document our passage in life's play but mediate our parts. We can interact with the data left by others alongside direct interactions with people themselves. The world, then, is not merely a stage but also a participatory exhibit.

The goal of this article is to clarify the ontological (rather than emic or phenomenological) distinction between actor and artifact. The actor performs in real time for an audience that monitors the actor. The artifact is the result of a past performance and lives on for others to view on their time. In making this distinction, I contend it is possible to extend current theories of online interaction and answer existing research questions such as: Why is it that contexts have "collapsed" online, as boyd suggests? Why is it so hard to nail down the notion of a friend online? How tightly can we couple the identity of an individual online and the activities of that individual? Addressing these questions entails a distinction between the sorts of online spaces where actors behave with each other ("performance" spaces, or behavior regions; Goffman, 1959) and "exhibition" spaces where individuals submit artifacts to show to each other. Clarifying

Bernie Hogan, "The Presentation of Self in the Age of Social Media: Distinguishing Performances and Exhibitions Online," *Bulletin of Science, Technology & Society*, vol. 30, no. 6, pp. 377–385. Copyright © 2010 by Sage Publications. Reprinted with permission.

this distinction creates an expanded theoretical repertoire for scholars, thereby enabling them to disentangle processes occurring when actors are copresent (in time, if not in the same geographic place) and processes that occur when actors are not necessarily present at the same time but still react to each other's data.

An exhibition is still a form of presentation of self. One can find off-line personal exhibitions in the presentation of photos in someone's house. Indeed, Halle (1996) indicates how class clearly differentiates the choice of artwork (or lack thereof) on display in living rooms. This is to say that people take their choice of what to display personally and consider it a form of impression management.

This distinction between performance and exhibition should be useful to scholars who are interested in the presentation of self online, and those who, like this author, consider notions of impression management a useful theoretical foil for understanding online behavior (boyd, 2007; Marwick & boyd, in press; Mendelson & Papacharissi, 2010; Lewis, Kaufman, & Christakis, 2008; Quan-Haase & Collins, 2008; Schroeder, 2002; Tufekci, 2008).

I begin this article with a review of Goffman's dramaturgical approach and its extensive use within social media studies. I then introduce the exhibitional approach, paying particular attention to the "curator," a key role generally absent from everyday life situations. In the penultimate section, I cover two areas of concern on social network sites (friend lists and collapsed contexts), where shifting the focus toward exhibitions may reveal new insights and facilitate future research agendas.

Goffman's Dramaturgy

Goffman's dramaturgical approach is a metaphorical technique used to explain how an individual presents an "idealized" rather than authentic version of herself. The metaphor considers life as a stage for activity. Individuals thus engage in performances, which Goffman (1959) defines as "activity of an individual which occurs during a period marked by his continuous presence before a particular set of observers and which has some influence on the observers" (p. 22). This continued presence allows individuals to tweak their behavior and selectively give and give off details, a process he termed "impression management."

One core assumption of the dramaturgical approach is that activity takes place in specific bounded settings. To explain this Goffman draws on Roger Barker's (1968) notion of the "behavior setting". In reacting to the behaviorism the early twentieth century (Skinner, 1939; Watson, 1913), Barker (1968) suggested most behavior was not determined by individual-specific stimulus-response patterns but was instead guided by the norms and goals of specific settings. Goffman (1959) distilled these specific settings into the well-worn dichotomy of the "front region" and the "back region," or more colloquially, the "front stage" and the "back stage." In the front stage, we are trying to present an idealized version of the self according to a specific role: to be an appropriate server, lecturer, audience member, and so forth. The backstage, as Goffman says, is "a place, relative to a given performance, where the impression fostered by the perfor-mance is knowingly contradicted as a matter of course" (p. 112). In the backstage, we do much of the real work necessary to keep up appearances.

What is key for this article is to highlight how situations are bounded in space and time. According to Goffman (1959),

> [W]hen a performance is given it is usually given in a highly bounded region, to which boundaries with respect to time are often added. The impression and understanding fostered by the performance will tend to saturate the region and the time span, so that any individual located in this space-time manifold will be in a position to observe the performance and be guided by the definition of this situation which the performance fosters. (p. 106)

This quote parallels Barker key qualities of a behavior setting (adapted from Heft, 2001, pp. 253–254):

- Specifiable geographical location
- Temporal boundaries
- Boundaries are perceptible
- Behavior settings exist independently of any single person's experience of them

Considering these qualities of the situation, Goffman's (1959) dramaturgical ap-proach is quite apt. Much like a stage play (rather than the script), it is bounded in space and time, and represents the instantiation of specific roles. Players seek to perform their role as convincingly as possible, and for the show to succeed there is much work that must take place behind the scenes. That these regions are bounded in time is implicit in how Goffman discusses shifts in performances:

> By proper scheduling of one's performances, it is possible not only to keep one's audiences separated from each other (by appearing before them in dif-ferent front regions or sequentially in the same region) but also to allow a few moments in between performances so as to extricate oneself psychologically and physically from one personal front, while taking on another. (p. 138)

The Audience

Within the dramaturgical approach, the audience refers those who observe a specific actor and monitor her performance. More succinctly, these are those for whom one "puts on a front." This front consists of the selective details that one presents in order to foster the desired impression alongside the unintentional details that are given off as part of the performance. Underlying this notion is the idea that the audience makes a single coherent demand on the individual. The above quote ("By proper scheduling …") reminds us that Goffman not only considers different regions as bounded in space-time, but that the audiences are bounded as well. That is to say, there is usually one specific front that needs to be presented in any given situation, because each region is not just a space-time locus, but a *time-space-identity* locus inhabited by a specific audience. Thus, it does not matter if the waiter knows his customers personally, so much that the waiter puts on that specific front to the customers. Moreover, a front involves the continual adjustment of self-presentation based on the presence of others. Goffman (1961) reinforces this idea in *Encounters*, by discussing unfocused and focused interactions. Focused interaction "occurs when people effectively agree to sustain for a time a single focus of cognitive and visual attention" (p. 7). But even in unfocused interaction such as "when two strangers across the room from each other … [each] modifies his own demeanor because he himself is under observation" (p. 7). The key point here is that individuals put on specific fronts and modify said fronts because of the sustained observation of an audience.

Goffman also notes that conflict can arise when fronts collide. In *The Presentation of Self in Everyday Life,* Goffman (1959) discusses the civil inattention that takes place when someone answers a telephone in front of others, or when conversations in public are loud enough to be heard by a third party. Similarly, Ling (2008) discusses the problem associated with the "dual-front" that emerges from the cell phone. He notes how an office phone that is tethered to the place of work represents the individual in that place, and is part of the rituals that constitute the office. In contrast, the cell phone connects people in many situations including ones where there is substantial mismatch between the two fronts (such as the high-powered business deal that gets done at the otherwise languid airport terminal).

Goffman as Applied to Online Media

Goffman might not consider himself a media scholar, although Lemert (1997) makes the case that Goffman is a product of the televisual age. And to the extent that he does, it might be more for *Frame Analysis* (Goffman, 1974) than for *The Presentation of Self in Everyday Life* (Goffman, 1959). Nevertheless, Goffman's dramaturgical approach is frequently considered a useful foil for understanding online presentation of self. The following list contains some of the many articles that use Goffman to this end:

- Donath (1998) employed Goffman as a starting point for signaling theory.
- Schroeder (2002) uses Goffman's dramaturgy quite literally in his analysis of virtual worlds.
- boyd (2004, 2006, 2007) used Goffman to ground SNS activity as networked identity performance.
- Hewitt and Forte (2006) use Goffman to explain identity production on Facebook and conflict because of the use of multiple fronts.
- Robinson (2007) argues for the effectiveness of Goffman's approach over postmodern perspectives found in Turkle (1997).

- Lewis, Kaufman, and Christakis (2008) draw on Goffman's front stage/back stage distinction for deriving research questions about privacy.
- Tufekci (2008) builds her research on Facebook presentation around Goffman alongside Dunbar's social brain hypothesis.
- Quan-Haase and Collins (2008) use impression management to discuss the art of creating status messages that signal availability.
- Menchik and Tian (2008) use Goffman and symbolic interactionism more broadly to interpret "face-saving" on e-mail mailing lists.
- Mendelson and Papacharissi (2010) demonstrate that pictures on social network sites conform to traditional notions of impression management.

A common thread running through these articles is that individuals would employ impression management (or the selective disclosure of personal details designed to present an idealized self). However, several articles draw more explicitly on the dramaturgical approach to suggest that sites based on access control are inherently private, and therefore, a "back stage" (boyd, 2006; Lewis et al., 2008; Robinson, 2007).

The notion that media provide a window into the private lives of others (or into things they would not normally show in public) is not specific to social media. This idea was used by Meyrowitz (1986) to explain some of the cultural impacts of television. He asserts that television exists in a private space and shows private lives: "through electronic media, groups lose exclusive access to aspects of their own back region, and they gain views of the back regions of other groups" (Meyrowitz, 1986, p. 135). Children get to view the typical adult world of their parents, men and women are now privy to conversations that would normally be segregated, and idols are brought down to earth through tabloid journalism. It is from here we can see the genesis of boyd's "collapsed contexts" (2007), as well as concerns about impression management vis-a-vis tabloid journalism and television's focus on scandal.

Backstage Is Not Private Space

I consider two issues emerging from this model: the conflation of the backstage with private spaces and the conflation of presentation of self with performance.

Several researchers have used the idea that Facebook is a backstage (Lewis et al., 2008; Tufekci, 2008) in order to motivate questions about privacy. However, the idea that some information is to be withheld from people is not the same thing as saying this information was part of what went into the creation of a front or that it contradicts a front as matter of course. From Goffman's definition, anywhere can be a back stage to another front stage. Academics working in their office present a front to the colleagues at their department by showing studiousness and perhaps not surfing the net. However, this front may also involve long periods of deliberation on a piece of work that is hidden from another front: the audience at a conference.

Online, the notion of a backstage fails to capture the role of a third party in regulating who has access to information about an individual. That Facebook allows only friends or "friends of friends" to see specific content does not suggest that this content signifies a backstage to other possible content that is available for anyone to see. To expect privacy online is not to imply that one has something worth hiding or a presentation that may contradict one's role in other spheres of life. Rather, it signifies that some individuals are classified as being considered contextually appropriate for this specific information (Nissenbaum, 2004). It further suggests that there is a third party (Facebook's servers) that knows who is considered an appropriate audience member for this content and who is not.

Lewis et al. (2008) used the notion of the backstage when comparing cultural information displayed by individuals with private and public accounts. They discovered that those with public accounts actually display more obscure music tastes. They loosely connected this to the notion of the backstage and suggest that some individuals draw open the stage's curtain to let the world see their tastes. To make their metaphor successful, they imply that music tastes are something inherently private and something that go into the creation of a front stage. However, it is more likely that showing music tastes is appropriate to the context of Facebook. Musical tastes are not a backstage but rather are a front. Some people carefully select which tastes to show, and thus, give a clear reason to make their profile less private. It is not that others with a narrower range of music want to hide their musical tastes but that they are indifferent to the association of taste and identity.

Artifacts Are Representations Not Performances

Beyond the issue of the back stage and privacy is a deeper issue about whether online content can be considered a performance in the first place. The conflation of performance and online profile is likely because of the notion that because a blog or profile signifies a single individual it does not merely stand in for that individual but is that individual (Reed, 2005). Similarly, Robinson (2007) coins the term *cyberperformers* to denote individuals who perform in cyberspace. In doing so, she equates the behavior of individuals in chat rooms and instant messengers (who either interact in real time or with specific known recipients) with the behavior of Flickr.com photo submitters and bloggers.

Can all content be considered a performance? To address this issue, it is useful to distinguish between performance as ephemeral act and performance as recorded act. Once a performance has been recorded, the nature of the performance has altered. It may still be a presentation of self, and undoubtedly it continues to signify an individual. However, it no longer necessarily bounds the specific audience who were present when the performance took place. Instead, it can be taken out of a situation and replayed in a completely different context. For example, a concert video may bring back great memories of a summertime show, but it does not transport the band to the viewer's living room.

The distinction between ephemeral act and recorded has an instructive parallel in the domain of art. In *The Work of Art in the Age of Mechanical Reproduction*, Benjamin

(1967) considers the functions of art in a time when process and reproduction make most artwork easily accessible to the masses. He asserts that these reproductions lack the unique "aura" of the original object. This aura is not a transcendental force but simply the unique historical trajectory of a singular object. This distinction between unique artwork with its aura and mechanical reproductions designed to signify the original parallels the distinction between singular individual, with one's own mind and presence, and digital traces designed to signify the individual.

Benjamin (1967) notes several consequences of this shift away from an emphasis on the aura of objects that also have a relevant parallel. First, in being reproduced, the reception of art becomes less something to be revered in a unique situation and more something to be consumed alongside other work:

> In the same way today, by the absolute emphasis on its exhibition value the work of art becomes a creation with entirely new functions, among which the one we are conscious of, the artistic function, later may be recognized as incidental. (p. 225)

Benjamin also suggests that individuals should be dissociated from their reproductions. All historically unique objects (including people) have an aura. He suggests that film is what separates the person from their aura:

> [F]or the first time-and this is the effect of film—man has to operate with his whole living person, yet forgoing its aura. For aura is tied to his presence; there can be no replica of it. (p. 229)

Third, he presages the difference between immediate impression management and the context-collapsing artifacts online. He does this by considering the fixed gaze of the camera:

> The film actor lacks the opportunity of the stage actor to adjust to the audience during his performance, since he does not present his performance to the audience in person ... The audience's identification with the actor is really an identification with the camera. (p. 228)

Thus, embedded in Benjamin's (1967) thesis about artwork is a relevant distinction between the individual and the representation of the individual. Benjamin, as well as those writing in his wake tended to focus on the consequences of art and film (cf. Hansen, 1987). However, there is nothing in his thesis that prevents us from importing these ideas into everyday life—now that everyday life is replete with reproductions of the self. To link this notion more explicitly to Goffmanian impression management, I offer below an explication of exhibition sites.

Exhibitional Approach Introduced

An exhibition site can now be defined as a site (typically online) where people submit reproducible artifacts (read: data). These artifacts are held in storehouses (databases). Curators (algorithms designed by site maintainers) selectively bring artifacts out of storage for particular audiences. The audience in these spaces consists of those who have and those who make use of access to the artifacts. This includes those who respond, those who lurk, and those who acknowledge or are likely to acknowledge.

Scope

In contrast to situations, many social media sites do not depend on being bounded in space and time with continued observation occurring between individuals. Instead they have the following features, which I consider sufficient components of an exhibition space:

1. Information signifying an individual is delivered to the audience, on demand by a third party.
2. Because of the reproducibility of content and the fact that it is sent to a third party for distribution, the submitter does not continually monitor these data as an audience is receiving it, and may possibly never fully know the audience.

Sites such as Facebook.com, Flickr.com, and YouTube.com have these qualities, as do the talk pages on Wikipedia.org (where content is associated with contributors). Wikipedia article pages would not be considered exhibition sites since the article is not designed to signify the specific individuals who wrote the article. Blogs generally fulfill these criteria but online gaming sites would not. That is to say, these criteria are most closely associated with what we presently consider social media or social network sites (boyd & Ellison, 2007).

The first and fundamental criterion draws a line between that which requires the *present* in order to be understood and that which makes no such demand. Virtual worlds and most online gaming (particular first-person shooters and MMOR-PGs) take place in the present. A user's actions are not simply placed in a sequence (such as reply-to), but are understood through mutual reactions where the timing of each action is relevant. Although individuals are not copresent in space, they are still monitoring and reacting to each other. The context of the game or the social world stands in for the context of a specific setting (Schroeder, 2002). In contrast, exhibition spaces require a third party to store data for later interaction; real-time interaction can take place, but it is not necessary. This is clearly unlike a "situation" as is noted by the aforementioned quotes from Goffman.

The second criterion draws a line between that which is *addressed* and that which is *submitted*. Some content is addressed to a particular person or some particular people. E-mail and instant messaging are examples of addressed media. Each message denotes a specific sender and a specific set of recipients. This is much like a situation where people are addressing specific alters or a specific audience. It is not necessarily in real time, but one can still put up a front intended for a specific set of recipients, and monitor activity in a direct reply. In exhibition spaces content is submitted to a data repository; people post status updates to Facebook, upload pictures to Picasa.com or Flickr.com, and post articles to a blog. This latter content may be produced and submitted with a specific audience in mind, but those who view and react to this content may be different from those for whom it was intended (if it was intended for anyone in particular to begin with).

These criteria do not preclude the use of an exhibitional approach in other domains but to suggest spaces where it is most appropriate: blog posts, photo galleries, and status updates. These are places where content is submitted to a third party, available to a large and potentially unknowable audience, and tethered to a specific submitter. The extension of the exhibitional approach to other spaces (and to hybrid spaces such as Google Wave) is beyond the scope of this article.

The Curator

Unique historical artifacts have typically been curated by experts. These people select which artworks to display, where to place them, and what narrative to tell about this selection. With a shift from presence (and aura) to data and reproduction, it is now possible for information signifying someone to be endlessly copied and reconfigured. Everyone can have his or her own exhibit, as long as the relevant information can be displayed with some coherence. Yet it is simply impractical to have a human curator pore over one's social information and devise a unique and relevant exhibit for each person, on demand. Consequently, computers have taken on this role, devising continually more sophisticated ways to curate artifacts.[1]

Curators mediate our experience of social information. Good curation presents things to the user that the user finds relevant or interesting. Bad curation is either overwhelming or unexpectedly irrelevant. Curators facilitate the following functions, which are available online and generally not a part of performances and situations: filtering, ordering, and searching. These functions are based on the fact that storehouses keep more artifacts than are generally on display. As such, it is necessary to limit the artifacts in some meaningful way.

Filtering artifacts simply limits which artifacts are on display. This can be done based on qualities of the artifacts or qualities of the relationship between an individual and the artifacts. For example, one might want to view only tweets that mention a specific topic. If the tweet is public and mentions the topic, it is included in the set of things to be displayed. If it is private, then the curator determines access to this tweet. If I am following someone's private account, then I can view these tweets.

Filtering performances is not something that can be done in a situation. Granted, one can choose to ignore a performance or specific aspects of it. People may choose to censor a humorous story for a specific group. But selectivity in a situation is not the same as filtering. Performers censor, curators filter on behalf of the audience. We can "tune in" or "tune out" performances, but filtering implies that one can evaluate a *set* of things *before* they are presented for consumption. Curators can do this because they retrieve things from a storehouse and put them on display.

Artifacts are also *ordered* in some way. Depending on the task, there is often a meaningful ordering. Communication is usually presented in reverse chronological order. Items for sale are frequently ordered according to price. More sophisticated algorithms can order items by relevance. For example, Facebook will select potential friends for the user from the larger set of friends. These potential friends will be ordered using a black box statistical metric seemingly related to the individuals one is likely to know. Amazon orders potential products based on their perceived relevance, which is a rank order based on a statistical measure of similarity. Lists of names are often sorted alphabetically.

Again, ordering is not something that can be done in a performance. That is not to say things in situations do not have a sequence. It is to say that performances in situations cannot be "reordered" as convenient. The order of online artifacts is based on the fact that each artifact is part of a set of similar artifacts that are known ahead of time. Performances have sequence but because they take place in "real time" or have a specific space-time locus they cannot be resorted at will.

Finally, artifacts online can be *searched.* Searching is simply filtering (and ordering) based on user input. Curators often work passively, as when people view their RSS readers, their Twitter queues, or their Facebook news feeds. However, sometimes filtering and ordering is done on content that includes specific requests from a user. Simply by viewing online content one is subject to filtering and ordering. Searching requires the user to submit additional information to fine-tune the display of content.

The role of the curator is to manage the preexisting content on behalf of the submitters. Within this space, it is more relevant to ask about the access controls that the curator put in place than whether or not this space is private. We may ask about the consequence of a specific ordering of data and whether this ordering is effective. We may also ask what is hidden from the users as a result of filtering, or what data are available for users to reorder. For example, can one reorder friends based on the number of mutual ties? Can one restrict access to content to a specific group of friends (i.e., impose a filter based on specific audience members)? How clearly do individuals understand different groups of friends on a given site? How easy is it to move content from one site to another?

Limits of the Exhibitional Approach

The exhibitional approach does not cover all online interaction, much like the dramaturgical approach does not cover all off-line interaction. For example, virtual worlds are hybrid spaces that share aspects of both off-line situations and online exhibitions. Insofar as there are servers that mediate information between individuals who are not immediately copresent, there is some recording involved. But play in social worlds generally takes place in specific bounded locations at specific bounded times in the same way that off-line interaction takes place in situations. One's avatar is meant to signify her mind as acting in a virtual context. The avatar interacts directly with other avatars that appear on the screen within one's field of view. It simulates off-line interaction, and consequently simulates the situation. Thus, it is unsurprising that Goffman has already been applied to these spaces (Schroeder, 2002).

Examples That Apply an Exhibitional Approach

In the penultimate section, I illustrate some examples where an exhibitional approach may illuminate or at least reorient our interpretations of online spaces.

What Is a Friend Online?

Sharing artifacts online is often done through "friends." As such, people add many friends to their online profile in order to participate in these sites fully. Curators use this list of friends in order to determine how to properly redistribute content. This list, however, is not tethered to a situation, but to an individual, beyond any specific situation. Consequently, people can add many more friends than would normally be included in a specific situation. It is not uncommon for students to have more than 200 friends on a social networking site (boyd, 2007; Lewis et al., 2008). This is larger than the number of people one is likely to know personally and feel close to. Depending on the question asked and method used, the number of people in the personal network varies from the low thrties (Hogan, Carrasco, and Wellman, 2007) through the upper sixties (Boase, Horrigan, Wellman, & Rainie, 2006; McCarty, Bernard, Killworth, Shelley, & Johnsen, 1997) to upwards of 150 (Roberts, Dunbar, Pollet, & Kuppens, 2008), but rarely if ever above that.

The irony of this situation is that only ten years ago, sociologists and those in related fields were actively assessing whether online interaction was isolating people (Kraut, Lundmark, Kiesler, Mukopadhyay, & Scherlis, 1998; Nie, Hillygus, & Erbring, 2002). Yet in 2009, the most recent OxIS report in Britain notes ex-users and nonusers of the Internet report twice as much of a sense of loneliness as Internet users (Dutton, Helsper, and Gerber, 2009). At the same time, people online are complaining the need to manage overwhelming lists of friends (Acquisti & Gross, 2006; Hewitt & Forte, 2006). This is unsurprising as there cognitive limits to the number of people one can actively maintain in a personal network (Dunbar, 1998).

If we consider online friends not as a means for signifying those with whom we have close relations but those with whom we want to manage access to content, we can refocus both what a friend means online and how to manage the surging lists of friends on many social network sites. How can systems be designed in order to curate more effectively? How do users classify their friends relative to a classification that emerges from the traces of interaction on a website? Gilbert and Karahalios (2009) approach this latter question by focusing on the ways in which strong ties can be modeled through passive data, such as time to last message or mutual friends. This work is oriented toward the ordering of content that has already been submitted. It is therefore possible to consider it as a means for fine-tuning the curatorial process. However, there is still little work on the means for fine-tuning the submission process. Do strong ties represent a single group to which one submits content? Or are there different strong ties within different groups whereby it is more useful to submit to the group and have group members filter accordingly? Here I do not provide an answer but reframe the question so that an answer can more effectively conform to the reality of what a "friend" is in an exhibition space.

Collapsed Contexts and the Lowest Common Denominator?

Friends are a form of access control online, and followers are a form of information management. These metaphors (friend and follower) do not perfectly correspond to their original meanings. Nevertheless, they are evidently a useful way to simplify the process of granting access controls online. If anything, these metaphors may be too simple. Boyd (2006) lists thirteen plausible reasons for befriending someone she encountered in ethnographic studies with teens. Only one was being a friend. The remainder focused on popularity, concerns for access control, and difficulty in saying no. What has emerged from this underdetermined friend tag is the accumulation of many social circles of friends under a single rubric (Hewitt & Forte, 2006). As sites expand to encompass more individuals from one's off-line life, with no clear distinction between them it also collapses all of the partially overlapping social circles of modern life (Simmel, 1922) into a single list. Friends may now refer to family members, coworkers, actual friends, neighbors, acquaintances, high school friends, people from online hobby groups or gaming sites, one-night stands, distant friends of friends, students past or present, and generally any other potentially personal relationship.

Boyd (2007) has referred to the existence of all of these groups in one space as the "collapsed contexts" quality of social network sites. For each of these contexts, one might have a slightly different presentation of self. Yet since they all have on-demand access to one's online artifacts, this results in a decontextualization of any of these artifacts. Artifacts are not tied to situations but to individual profiles. The individual therefore comes to represent these same artifacts to all "friends." If social network sites house more friends than are cognitively manageable, all of whom have access to one's content, and many of whom represent different social groupings and different potential fronts, then how do individuals manage to submit any content at all? Why is there not a sense of self-presentation paralysis?

The answer is that one need not consider everyone when submitting content but only two groups: those for whom we seek to present an idealized front and those who may find this front problematic. That is, in addition to the traditional audience of situations, one must add a hidden audience who are not the intended recipient of content but will have access to it as well. One might not post for one's boss on Twitter, but if one's boss is following (or is likely to follow), then one will certainly post in light of the fact that the boss may read it. One might not be posting for one's parents (or children or students) on Facebook, but again, one is posting in light of the fact that these individuals may have access; these individuals define the *lowest common denominator* of what is normatively acceptable.[2]

A theory of lowest common denominator culture is more appropriate to exhibition spaces replete with persistent content than single context performances. It offers a potential explanation for three aspects of social network sites. The first is why individuals effectively participate in these sites, halfheartedly join, or even refuse: An individual assesses whether his identity can be effectively represented by the lowest common denominator of the people who view his content in his absence. The second is to explain how in an age of profound surveillance (both from authorities and peers), individuals still submit content that is unambiguously questionable (nudity, violence, political extremism, racial epithets): The lowest common denominator of niche sites may be different than that of general sites. As such, one may have a clean profile on Facebook but a series of lewd pictures on Xtube.com, Suicidegirls.com, Pornotube.com, and so forth. Similarly, one may be sexually ambiguous or even deceptive on Facebook or one's Twitter account, but still have a openly gay profile on Gaydar.co.uk, Gay.com, Manhunt.com, and so on. A businessperson may seek to be clean cut and professional on one site but espouse politically extreme views on Stormfront.org or Newsaxon.org. In more positive terms, a teacher may complain about troubling students on TheApple.Monster.com but make no such claims on Facebook, where the teacher might be friends with students' parents or the students themselves. The third aspect of these sites that a theory of lowest common denominator addresses is how exactly individuals interpret this particular context: it is likely that people do not create sophisticated projections of their social network, nor need they. Instead, their behavior is in reference to specific salient individuals, who are small enough in number to be coherent. The persistence of this content beyond these salient individuals is rarely accounted for. This theory is also in keeping with research by Acquisti and Gross (2006) about why individuals will reveal a great deal of information on Facebook: they trust the site to curate it for them appropriately (even though they are often misinformed about the who can access what), and that they submit information they feel is inoffensive to some perceived salient individuals.

Conclusions

Many online sites set up a situation where individuals can continually submit data to be associated with their profile. This sort of "interaction" where people view and react to the submitted content of others is dissimilar from the traditional situations that gave rise to Goffman's germane dramaturgical approach. The impetus for this article was to suggest that many aspects of Goffman's approach (e.g., impression management) can work in a framework that is more aligned to these spaces, namely through the metaphor of an exhibition rather than one of a stage play. One of the key distinctions between exhibitions and performances is that performances are subject to continual observation and self-monitoring as the means for impression management, whereas exhibitions are subject to selective contributions and the role of a third party. I refer to this third party as a curator that has the capacity to filter, order, and search content. The exhibition has its own logic, such as lowest common denominator culture and easy persistent friends that do not have direct analogs in offline life. Privacy becomes a matter of how content is aggregated (e.g., Is it tied to a real name? Is it tied to a geographical location? How findable is the user and the user's data?) and how access control is managed *a priori,* rather than *in situ.* I draw on Benjamin's classic essay to indicate that the notion of distinguishing individuals from their reproductions is not a novel idea. What is novel, however, are its application to everyday life, and its aggregation through digital means.

Acknowledging the difference between performances and exhibitions is an ontological matter, not an empirical or phenomenological one. People need not acknowledge this mediated relationship in order to participate in it. In fact, it is likely that ignorance blissfully facilitates the willing capture, storage, and use of private data. Moreover, it is a difference that allows individuals to consume and view each other's past artifacts without directly engaging the individual, or in many instances, even letting the profile owner know that their information is being viewed.

What is empirical in this domain is the extent to which misunderstandings about the basic ontological structure of data, its curation, and exhibition give rise to new unintended problems: social information overload, collapsed contexts, accidental disclosures, and "identity" theft. What people do is based on their mental models of these sites, and as past work has demonstrated, there is often a great mismatch between the mental models and the actual behaviors (Acquisti & Gross, 2006). These sites also give rise to new potentials: heightened social capital from newly accessible weak ties (Ellison, Steinfeld, & Lampe, 2007); asynchronous and coordinated grassroots organizing (Earl, 2010); strengthening of long distance ties through photo sharing and rapid exchanges (Cook, Teasley, & Ackerman, 2009); otherwise hidden communities such as gays and lesbians suddenly being able to find each other rapidly and privately (Gray, 2009). These new phenomena are not necessarily about the performance but the new mediated architecture that encapsulates and redistributes past performances for mutual and often asynchronous benefits. The capabilities of exhibition sites allow a person to be found when others want to look rather than when the person is able to be present and perform. Thus, extending presentation of self by considering an exhibitional approach alongside a dramaturgical one is meant to be a step toward a clearer articulation of both the potentials and the perils of self-presentation in an age of digital reproduction.

Acknowledgments

The author would like to thank the anonymous reviewers along with danah boyd, Scott Golder, Allison Powell, Ralph Schroeder, Amaru Villanueva Rance, and the participants of the Oxford Internet Institute's 2009 summer doctoral program for their input and constructive criticism.

Notes

1. Interestingly, "digital curation" is now used in the United Kingdom to refer to the practice of maintaining storehouses of digital research content (see http://www.dcc.ac.uk/). But in this case, it is still implied that there is an expert individual who is maintaining the data. This is a different matter, and one that parallels offline archival. This sort of curation also tends to work at the level of the data set, much like offline curation works with a specific artwork, rather than the level of raw data.

2. I would like to thank an anonymous reviewer for noting that lowest common denominator culture is not specific to exhibition spaces but to collapsed contexts. For example, a wedding speech might not cater to every single audience member but simply be inoffensive to salient individuals (e.g., a priest and one's mother-in-law) while appealing to friends and certain relatives. That said, even in the case of a wedding speech, certain poor jokes can immediately be "recovered" in ways that artifacts may not.

I Tweet Honestly, I Tweet Passionately

Twitter Users, Context Collapse, and the Imagined Audience

By Alice E. Marwick and danah boyd

Introduction

Wergence e present ourselves differently based on who we are talking to and where the conversation takes place—social contexts like a job interview, trivia night at a bar, or dinner with a partner differ in their norms and expectations. The same goes for socializing online.

Participants have a sense of audience in every mediated conversation, whether on instant messenger or through blog comments. This audience is often imagined and constructed by an individual in order to present themselves appropriately, based on technological affordances and immediate social context. Studies of identity presentation on profile-based sites, such as social network sites and personal ads, have demonstrated that profile owners are attentive to audience (e.g. boyd, 2006b; Ellison et al., 2006). The need for variable self-presentation is complicated by increasingly mainstream social media technologies that collapse multiple contexts and bring together commonly distinct audiences. This paper examines how people using the microblogging site Twitter imagine their audiences and what strategies they use to navigate networked audiences. Our findings shed light on how audience changes in networked environments.

Imagining the Audience Online

Every participant in a communicative act has an *imagined audience*. Audiences are not discrete; when we talk, we think we are speaking only to the people in front of us or on the other end of the telephone, but this is in many ways a fantasy. (Social norms against eavesdropping show how 'privacy' requires the participation of bystanders.) Technology complicates our metaphors of space and place, including the belief that audiences are separate from each other. We may understand that the Twitter or Facebook audience is potentially limitless, but we often act as if it were bounded. Our understanding of the social media audience is limited. While anyone can potentially read or view a digital artifact, we need a more specific conception of audience than 'anyone' to choose the language, cultural referents, style, and so on that comprise online identity presentation. In the absence of certain knowledge about audience, participants take cues from the social media environment to imagine the community (boyd, 2007: 131). This, the imagined audience, might be entirely different from the actual readers of a profile, blog post, or tweet.

Joshua Meyrowitz's germinal work *No Sense of Place* (1985) applied situationist theory to the changes brought about by electronic media like television and radio. Situationism maintains that people react to situations based on context rather than fixed psychological traits. Situations, and thus social order, are collectively produced by participants (Garfinkel, 1967). Meyrowitz theorized that electronic media eliminated walls between separate social situations, contributing to the rapid social change that characterized the United States in the 1960s. Similarly, self-presentation theory has been used to understand the further combination of audience by digital media. Self-conscious identity performances have been analyzed in internet spaces like social network sites (boyd, 2007; Livingstone, 2005), blogs (Hodkinson and Lincoln, 2008; Reed, 2005), dating sites (Ellison et al., 2006) and personal homepages (Papacharissi, 2002; Schau and Gilly, 2003).

Personal homepages, arguably the first multi-media online identity presentations, are highly managed and limited in collaborative scope; people tend to present themselves in fixed, singular, and self-conscious ways. Papacharissi describes the personal homepage as 'a carefully controlled performance through which self presentation is achieved under optimal conditions' (2002: 644). Schau and Gilly found that personal homepage creators thought of their work as constructed for the public; even if they focused on friends or family, creators 'acknowledge(d) the potential for the audience to be unlimited and undefined' (2003: 391). Similarly, Robinson argues 'the very construction of the homepage presumes the expectation of the virtual "generalized other"' (2007: 104). She notes that personal homepages are more atomized and isolated than blogs and message boards, which presume ongoing communication with the audience. She writes, 'the "I" is constantly redefined as the "me" in response to this interactional commentary' (2007: 104).

The specifics of the imagined audience are more important in social media that involve greater interaction with readers than personal homepages. Professional writers' sense of 'audience awareness' factors greatly into their writing, in terms of goals, vocabulary, technique, and subject matter (Berkenkotter, 1981). Like many writers, bloggers write for a 'cognitively constructed' audience, an imagined group of readers who may not actually read the blog (boyd, 2006a). Actual readers are present only in digital ephemera like server logs or comments. The imagined audience exists only as it is written into the text, through stylistic and linguistic choices (Scheidt, 2006). Through the process of labeling connections as 'Friends', social network sites require participants to publicly articulate connections, thereby enabling them to write their audience into being (boyd, 2006b).

In goal-oriented spaces like dating sites, people are highly conscious of audience. Ellison et al. (2006) found that personal ads were constructed with a hyper-aware self-consciousness, as users knew that misspellings, cultural references, and even time stamps were likely to be scrutinized by potential suitors. Similarly, social network site users select 'markers of cool' based on an imagined audience of friends and peers. Liu's (2007) study of 'taste cultures' on social network site profiles found that participants listed favorite books, music, movies, and TV shows to construct elaborate taste performances, primarily to convey prestige, uniqueness, or aesthetic preference.

The microblogging site Twitter affords dynamic, interactive identity presentation to unknown audiences. Self-presentation on Twitter takes place through ongoing 'tweets' and conversations with others, rather than static profiles. It is primarily textual, not visual. The potential diversity of readership on Twitter ruptures the ability to vary self-presentation based on audience, and thus manage discrete impressions.

Twitter

Twitter is a microblogging site, originally developed for mobile phones, designed to let people post short, 140-character text updates or 'tweets' to a network of others. Twitter prompts users to answer the question 'What are you doing?', creating a constantly-updated timeline, or stream, of short messages that range from humor and musings on life to links and breaking news. Twitter has a directed friendship model: participants choose Twitter accounts to 'follow' in their stream, and they each have their own group of 'followers'. There is no technical requirement of reciprocity, and often, no social expectation of such. Tweets can be posted and read on the web, through SMS, or via third-party clients written for desktop computers, smartphones, and other devices. These different access methods allow for instant postings of photos, on-the-ground reports, and quick replies to other users. The site was launched in 2006, and broke into the mainstream in 2008–09, when accounts and media attention grew exponentially. In May 2009, the Nielsen Company reported that Twitter had approximately 18.2 million users, a growth rate of 1,448 percent from May 2008. Today, the most followed Twitter accounts represent public figures and celebrities, from US president Barack Obama to actor Ashton Kutcher and pop star Britney Spears.

Twitter and Audience

As in much computer-mediated communication, a tweet's actual readers differ from its producer's imagined audience. Twitter allows individuals to send private messages to people they follow through direct messages (DMs), but the dominant communication practices are public. A convention known as the '@reply' (consisting of the @ sign and username) lets users target a conversation to or reference a particular user, but these tweets can be viewed by anyone through search.twitter.com, the public timeline, or the sender's Twitter page (Honeycutt and Herring, 2009 provide a detailed discussion of @ replies).

On Twitter, there is a disconnect between followers and followed. For instance, musician John Mayer (johncmayer) is followed by 1,226,844 users, but follows only 47. While followers provide an indication of audience, this is imprecise. When an individual's account is public, anyone—with or without a Twitter account—can read their tweets through the site, RSS, or third-party software. The vast majority of Twitter accounts are public. Those who choose to protect their accounts can restrict their audience, but the lists of followers on both public and protected accounts indicate only a potential audience, since not everyone who follows a user reads all their tweets.

Tweets are also spread further when participants repost tweets through their accounts. This practice, commonly referred to as 'retweeting', can introduce content to new audiences (boyd et al., 2010). While the dominant norm is to use @username to cite the original author or attribute the person who spread the message, retweeted messages are often altered and may lose any reference to the original. Additionally, it is not uncommon for people to forward tweets via email or by copying and pasting them into new communication channels. Furthermore, various tools allow users to repost tweets to Facebook, MySpace, and blogs.

Given the various ways people can consume and spread tweets, it is virtually impossible for Twitter users to account for their potential audience, let alone actual readers. Yet, this inability to know the exact audience does not mean that tweets are seen by infinite numbers of people. As with blogs (Shirky, 2005), nearly all tweets are read by relatively few people—but most Twitterers don't know *which* few people. Without knowing the audience, participants imagine it.

How Twitter Users Imagine Audiences

Methodology

To find out how Twitter users imagine their audience, we asked them directly. We posted questions to our own followers (many of whom retweeted our question to their followers) and sent @reply questions to a sample of users whose tweets appear in the public timeline, every person in the 300 most-followed accounts (249 total), and a subset of users with 1,000–15,000 followers. While all the Twitter accounts we reference are public, we anonymized all of our informants except the highly-followed users. Our questions included: 'Who do you imagine reading your tweets?' and 'Who do you tweet to?' Later, we asked: 'What makes an individual seem "authentic" on Twitter? (Or what does it mean to be authentic?)' and 'What won't you tweet about? What subjects are inappropriate for Twitter?' Our goal in approaching different types of users was not to get a representative sample of Twitter users but to elicit potentially diverse perspectives.

Given the issues mentioned above, we are unable to assess how many people saw our tweets. We received 226 responses from 181 Twitter users through direct messages or @

replies to our queries. The responses we received revealed many different perspectives on audience. While we can neither quantify audience management techniques nor account for all potential perspectives, the responses we received provide valuable insight into some core differences in conceptualizing audience. Future empirical work might examine the prevalence of these strategies and their relationship to follower counts, demographics, or genres of content.

To Whom Am I Speaking?

Our informants conceptualized their audience on Twitter in diverse and varied ways. Most responses we received focused on abstract categories of people (e.g. 'friends'), but a few indicated that their audience was articulated through the service itself. For example, an informant defined his audience as 'the overlap between my followers and my following'.

Respondents with relatively few followers typically spoke about friends, but some focused on themselves. Respondents with large followings commonly described their audience as 'fans'. Of course, some have multiple audiences in mind:

> I think I write to the people I follow and have twittered something recently.
> And I also tweet to myself. Is that wrong?
> I guess I'm tweeting to my friends, fans … and talking to myself.

Although some respondents emphasized that they speak to friends through Twitter, what they mean varies. Part of the difficulty is that 'friends' is an overloaded term in social media (boyd, 2008). One user described her friends as people she followed, while another talked about writing to her 'IRL friends' to signal people she knew outside of Twitter. Such users imagined their audience as people they already knew, conceptualizing Twitter as a *social* space where they could communicate with pre-existing friends. This follows the argument that Twitter's strength is in its encouragement of 'digital intimacy' (Thompson, 2008). Many tweets are phatic in nature (Miller, 2008) and serve a social function, reinforcing connections and maintaining social bonds (Crawford, 2009). One respondent wrote, 'I guess it's like a live diary to all my friends. I post what they might find interesting or know they will have an opinion on.'

When respondents referred to their audience as 'me', they also meant different things. Some thought of Twitter as a diary or record of their lives. Others saw the service as a space where they expressed opinions for themselves rather than others. Emphasizing 'me' may also be a self-conscious, public rejection of audience:

> Myself. It is MY Twitter account so, it's mostly about me.
> < Who do you tweet *to*?> No one & I love that. Or maybe myself five min.
> ago: I write the tweets I want to read.
> I don't tweet to anybody; I just do it to do it.

Although these individuals may not direct tweets to others, they are not tweeting into a void; they all have followers and follow others. Their emphasis on 'me' implies that for them, Twitter is personal space where other people's reactions do not matter. Similarly, a few people saw crafting tweets for a particular audience as problematic:

> As an individual (not org or corp) it's worth it 2 me 2 lose followers 2 maintain the wholeness/ integrity of who/ what/how I tweet.
> when I tweet, I tweet honestly, I tweet passionately. Pure expression of my heart.

What emerges here is not that these individuals lack an audience, but that they are uncomfortable labeling interlocutors and witnesses as an 'audience'. In bristling over the notion of audience, they are likely rejecting a popularly discussed act of 'personal branding' as running counter to what they value: *authenticity*. In other words, consciously speaking to an audience is perceived as inauthentic.

The strategic use of Twitter to maintain followers, or to create and market a 'personal brand', is part of a larger social phenomenon of using social media instrumentally for self-conscious commodification. In this process, strategically appealing to followers becomes a carefully calculated way to market oneself as a commodity in response to employment uncertainty (Hearn, 2008; Lair et al., 2005). As Dan Schawbel, author of *Me 2.0: Build a Powerful Brand to Achieve Career Success*, writes on the blog Mashable:

> Today, Twitter has roughly 6 million users and is projected to grow to 18.1 million users by 2010. With all those people, the chances for networking are endless and connecting with new people can lead to career opportunities, so it is essential that your personal brand exists on the service … By leveraging the Twitter platform to build your brand you can showcase yourself to a huge and growing audience. (2009)

Using Twitter to carefully construct a 'meta-narrative and meta-image of self' (Hearn, 2008) is part of what Jodi Dean (2002) calls the 'ideology of publicity', in which we value whatever grabs the public's attention. Publicity culture prizes social skills that encourage performance (Sternberg, 1998); people are rewarded with jobs, dates, and attention for displaying themselves in an easily-consumed public way using tropes of consumer culture. In contrast, tweeting for oneself suggests a true-to-self authenticity, untainted by expectations. Of course, authenticity is a social construct (Grazian, 2003) and it is unlikely that anyone could tweet context-independently with no concern with audience, given our understanding of audience influence on self-presentation (Blumer, 1962; Goffman, 1959). We are interested not in an absolute sense of authenticity, but in what Twitter users consider 'authentic'.

Other respondents suggested that audience conceptions were tweet-dependent. From this perspective, Twitter is a medium, like telephony or email, that can be used for many different purposes:

isn't tweeting (like all things) situational? Try replacing the word 'tweet' in that Q [with] 'email'.
Q->A: depends on the topic& intensity of connection: a)RT for public b)@/ DM for followers/friends, c) "thinking aloud for myself";-)

This implies that users write different tweets to target different people (e.g. audiences). This approach acknowledges multiplicity, but rather than creating entirely separate, discrete audiences through the use of multiple identities or accounts, users address multiple audiences through a single account, conscious of potential overlap among their audiences. However, the difference between Twitter and email is that the latter is primarily a directed technology with people pushing content to persons listed in the 'To:' field, while tweets are made available for interested individuals to pull on demand. The typical email has an articulated audience, while the typical tweet does not. Email is also usually private, while Twitter is primarily public. Notably, people avoid broaching many topics on Twitter precisely for this reason.

Another approach some respondents took was to conceptualize their audience as an 'ideal' person:

> I imagine my audience as a fellow nerd, who gets a say in my amusement, confusion, or disappointment at whatever just happened.
> In super-deep novel-writing mode tho I doubt it looks like it, so can't talk, but basically I imagine a few "ideal readers".
> I think of a room filled with friends when I tweet. I assume people like me that are reading my tweets.

The 'ideal' reader is a well-known concept for writers, who often write 'to' an imaginary interested party. In the tweets above, the ideal reader is conceptualized as someone similar to the writer, who will presumably share their perspective and appreciate their work. Ethnographies of television production have similarly found that producers 'imagine others like and unlike themselves, (re)constructing their own identities in the process of constructing the imagined audience' (Peterson, 2003: 161). The *ideal* audience is often the mirror-image of the user.

Strategic Audiences

In contrast to general notions of imagined audiences, Twitter users with numerous followers expressed specific, pragmatic understandings of audience. A few mentioned real-life friends, family, and co-workers, but others with 100,000+ followers suggested that they imagined their audience as a fan base or community with whom they could connect or manage:

> Adventuregirl: when I tweet I thnk of all my Sweet Tweets and sharing my life's advenutes w/them- and luv hearing there's 2!

Padmasree: I mostly think of the community (more than friends)

This approach can be understood through the lens of 'micro-celebrity'. Senft describes micro-celebrity as a communicative technique that 'involves people "amping up" their popularity over the Web using techniques like video, blogs, and social networking sites' (2008: 25). Micro-celebrity implies that all individuals have an audience that they can strategically maintain through ongoing communication and interaction. Twitter is used this way by many people—including marketers, technologists, and individuals seeking wide attention—to establish a presence online. Likewise, by embracing social media to engage directly with their audience, many traditional celebrities and public officials embrace the techniques of micro-celebrity.

Users with large numbers of followers reasonably conceptualized and navigated their audiences tactically. For example, Casey Wright (100,000 followers) answered that he assumed 'a broad audience with disparate tastes.' When asked if he tailored tweets to different parts of the audience, he answered, 'Not really. I don't think any tweet reaches everyone but they all appeal to someone. I try to mix it up.' The specificity of audience understanding was striking among some with large followings:

> Nansen: I think of 1. political messaging 2. new friends 3. information 4. news
> GuysReplies: My tweets are news broadcasts ala NYTimes or StumbleUpon with Alltop plugs.
> Brooksbayne: all of the above, but i have different "silos" for convos here. hashtags help with all that.
> [Authors]: What are your different silos?
> Brooksbayne: politics, foodie, tech, social media, music biz, brands, and bacon!

Nansen, a conservative activist with 110,000 followers, has definite goals for Twitter: maintain consistent political messaging, create relationships with new friends, provide information, and spread news. Guy Kawasaki (154,000 followers) views his feed primarily as a news broadcast. Brooks Bayne (95,000 followers) recognized the diversity of his audience and used hashtags (keywords preceded by the # sign) to direct tweets to interested followers. These users were personally invested in maintaining high follower numbers and used several techniques to attract attention.

That is not to suggest there is an absolute divide in practice between the heavily followed and those who are not. Instead, knowledge of the audience functions more as a continuum. Several highly followed users did not mention trying to build and maintain audience or feigned unawareness:

> TychoBrahe: Honestly, I have no idea who reads them. Hopefully a very small group of very forgiving people!

Others acknowledged their visibility but didn't see their actual audience as their intended audience. Jason Goldman, a Twitter employee, said 'I sometimes think about what my girlfriend or coworkers or mom would think. I don't think about "audience" really ... If I think about audience before tweeting (mostly not true) I think "would my friends dig this."' Movie blogger Harry Knowles said, 'I imagine that my friends are reading mostly, but with the knowledge there's a greater voyeuristic society tuned in.' Of course, just because highly followed users claim that they're not focused on audience does not mean they are not. They may also be aware of the value of being perceived as authentic.

We also talked to users with fewer followers who had strategic plans for their audiences. In choosing what to put forward, they often learn to present what is well received.

> like my stream 1/3 humors, 1/3 informative, 1/3 genial and unfiltered, transparency is so chic. try to tweet the same way.
> Who do I imagine when I tweet? I think, I imagine myself getting my opinion out to hundreds of ppl who might care:-)
> U know I don't know who reads them, but when I tweet sumthin contrvsial or interesting I find a get a cple more followers.

The strategies of micro-celebrity are not only used by people with large numbers of followers. Many users consciously use Twitter as a platform to obtain and maintain attention, by targeting tweets towards their perceived audience's interest and balancing different topic areas.

A variety of imagined audiences stems from the diverse ways Twitter is used: as a broadcast medium, marketing channel, diary, social platform, and news source. It is a heavily-appropriated technology, which participants contextualize differently and use with diverse networks. The networked audience is an abstract concept and varies among Twitter users, in part because it is so difficult to ascertain who is actually there.

Navigating Multiple Audiences

The Need to Navigate

Like many social network sites, Twitter flattens multiple audiences into one—a phenomenon known as 'context collapse'. The requirement to present a verifiable, singular identity makes it impossible to differ self-presentation strategies, creating tension as diverse groups of people flock to social network sites (boyd, 2008). Privacy settings alone do not address this; even with private accounts that only certain people can read, participants must contend with groups of people they do not normally bring together, such as acquaintances, friends, co-workers, and family. To navigate these tensions, social network site users adopt a variety of tactics, such as using multiple accounts, pseudonyms, and nicknames, and creating 'fakesters' to obscure their real identities (Marwick, 2005). The large audiences for sites like Facebook or MySpace may create a lowest-common denominator effect, as individuals only post things they believe their broadest group of acquaintances will find non-offensive. Similarly, Twitter users negotiate multiple, overlapping audiences by strategically concealing information, targeting tweets to different audiences and attempting to portray both an authentic self and an interesting personality.

But why do users need to navigate multiplicity? In his seminal text *The Presentation of Self in Everyday Life* (1959), Erving Goffman conceptualized identity as a continual performance. Goffman analyzed people's practices using a dramaturgical metaphor, suggesting that we can understand individuals as actors who tailor self-presentation based on context and audience. He proposed that in any given situation, people, like actors, navigate 'frontstage' and 'backstage' areas; a workplace's office space and meeting rooms might be frontstage, while more candid talk takes place backstage at after-work happy hour. Goffman's work is often grouped with symbolic interactionism, a sociological perspective which holds that meaning is constructed through language, interaction, and interpretation (Blumer, 1962; Strauss, 1993). Symbolic interactionism claims that identity and self are constituted through constant interactions with others—primarily, talk. In other words, self-presentation is *collaborative*. Individuals work together to uphold preferred self-images of themselves and their conversation partners, through strategies like maintaining (or 'saving') face, collectively encouraging social norms, or negotiating power differentials and disagreements.

Goffman maintained that this becomes a process of 'impression management', where individuals habitually monitor how people respond to them when presenting themselves. This process is self-conscious in situations of intense scrutiny, like first dates and job interviews, but is habitual even in relaxed social situations. Self-monitoring leads people to emphasize or de-emphasize certain things, responding to further feedback in a dynamic, recursive process (Leary and Kowalski, 1990: 43). Thus, self-presentation changes based on audience factors, such as friendship ties (Tice et al., 1995), status differentials (Leary and Kowalski, 1990: 38), and racial differences (Fleming and Rudman, 1993). Even in difficult circumstances, people are skilled at using gesture, language, and tone to manage impressions face-to-face (Banaji and Prentice, 1994).

Most of these studies draw from data and observations that involve people interacting face-to-face, where it is fairly easy to gauge the gender, race, status, etc. of the audience. Removing this ability creates tensions. Meyrowitz (1985) gives the example of Black Power advocate Stokely Carmichael, who typically used different styles when presenting to black and white audiences. Speaking on broadcast television, Carmichael could not appear 'authentic' to both audiences and had to choose between a black or white rhetorical style. He chose the former, engaging his black audience but alienating white viewers. In today's media-saturated landscape, politicians and celebrities use 'polysemy' or coded communication to simultaneously appeal to different, even oppositional audiences (Albertson, 2006; Fiske, 1989). Madonna's early image exemplifies polysemy. She was interpreted differently by young women, who responded to her feminist message, and young men, who responded to her sexy persona (Fiske, 1989). Similarly, George W. Bush sprinkled coded references to hymns, Bible verses, and Evangelical culture throughout his speeches to appeal to his base without alienating others (Albertson, 2006).

Social media thus combines elements of broadcast media and face-to-face communication. Like broadcast television, social media collapse diverse social contexts into one, making it difficult for people to engage in the complex negotiations needed to vary identity presentation, manage impressions, and save face. But unlike broadcast television, social media users are not professional image-makers, and rather than giving a speech on television, they are often corresponding with friends and family. By necessity, Twitter users maintain impressions by balancing personal/public information, avoiding certain topics, and maintaining authenticity.

Balancing Expectations of Authenticity

The imagined audience affects how people tweet. People with few followers, who use the site for reasons other than self-promotion, generally see Twitter as a *personal* space where spam, advertising, and marketing are unwelcome. Following the paradigm of symbolic interactionism, identity on Twitter is constructed through conversations with others. Tweets are formulated based partially on a social context constructed from the tweets of people one follows. Participants must maintain equilibrium between a contextual social norm of personal authenticity that encourages information-sharing and phatic communication (the oft-cited 'what I had for breakfast' example) with the need to keep information private, or at least concealed from certain audiences. The tension between revealing and concealing usually errs on the side of concealing on Twitter, but even users who do not post anything scandalous must formulate tweets and choose discussion topics based on imagined audience judgment. This consciousness implies an ongoing front-stage identity performance that balances the desire to maintain positive impressions with the need to seem true or authentic to others.

This concept of 'authenticity' is a popular one. We refer to the 'real me' and authentic experiences, artifacts, and people. However, there is no such thing as universal authenticity; rather, the authentic is a localized, temporally situated social construct that varies widely based on community. Grazian's study of blues bars in Chicago defines authenticity as conforming to an idealized representation of reality. The authentic is always manufactured, and always constructed in 'contradistinction to something else' (2003: 13). In other words, for something to be deemed authentic, something else must be inauthentic. However, this dichotomy is false when we note that both the performance of authenticity and inauthenticity are equally constructed by discourse and context (Cheng, 2004). What we consider authentic constantly changes, and what symbols or signifiers mark a thing as authentic or inauthentic differ contextually.

The fact that we constantly vary self-presentation based on audience reveals authenticity as a construct: are we more or less authentic with our book club or gym partner? Whether we are viewed as authentic depends on the definition imposed by the person doing the judging. As in much social media, participants' understanding of authenticity on Twitter varies. For fashion bloggers, the ability to assemble an outfit that reflects a personal aesthetic and knowledge of larger trends marks one as authentically stylish and fashionable. For open-source geeks, on the other hand, ignorance of current trends marks authenticity, emphasizing instead mental acuity and knowledge of software and information law. Since authenticity is constituted by the audience, context collapse problematizes the individual's ability to shift between these selves and come off as authentic or fake. We observed Twitter users using two techniques to navigate these tensions: self-censorship and balance. People refrain altogether from discussing certain topics on Twitter, while others balance strategically targeted tweets with personal information.

Self-Censorship

Some people we spoke with suggested that they simply would not broach certain topics on Twitter. Self-censorship can be a useful technique in the face of an imagined audience that includes parents, employers, and significant others. Some respondents assumed anyone could potentially read their tweets, making it impossible to discuss controversial or personal topics:

> anything i'd consider TMI (to spare my followers): family problems, relationship rants, etc. This ain't FB.
> bathroom activity, romantic relationships, complaining about an employer
> Won't Tweet anything too personal, TMI about self/others, dead horse areas from subjects like religion/politics/sports

Subjects mentioned included dating, sex, relationships, marital problems; Too Much Information (TMI) about bodily functions and the like; criticism of one's job; and controversial or negative topics that might alienate followers. Without the ability to vary information flow based on audience, participants could not risk a sensitive topic being viewed by the wrong person.

Twitter can be viewed as a public space that should be carefully policed:

> i'm very conscious that twitter is public. i wouldn't tweet anything i didn't want my mother/employer/professor to see
> I tweet about anything I would say in a lobby. Beyond that, each tweet is influenced by the tweets around that unique moment.

Interestingly, one user views Facebook as an appropriate place for 'family problems or relationship rants'. While many respondents self-censored sex and dating information, these topics are often discussed on blogs, Facebook, and LiveJournal. The information was not too sensitive to ever be revealed, but it could not be revealed *on Twitter,* which was seen by some as a 'professional' environment with potential professional costs:

> I think it all depends on what the intended purpose for your twitter account. Professionals should beware how they rep their cos i got threatened w/ lawsuit and loss of work bc of one of my tweets. quite careful now in what i tweet. or try to be!

Work concerns influenced what people tweeted about as well as what they self-censored. For instance, a freelancer said 'I always keep my clients in mind. I want to convey intelligence and professionalism, and diversity—I want to be seen as interested in a lot of things.' She could present herself appropriately against the social context of an imagined audience of other professionals.

Participants maintain a public-facing persona to manage impressions with potential readers. Context collapse creates an audience that is often imagined as its most sensitive members: parents, partners, and bosses. This 'nightmare reader' is the opposite of the ideal reader, and may limit personal discourse on Twitter, since the lowest-common-denominator philosophy of sharing limits users to topics that are safe for all possible readers. While people do talk about controversial subjects on Twitter, our respondents show that some are more likely to avoid personal topics that imply true intimacy and connection between followers. Instead, they may frame Twitter as a place where the strictest standards apply.

Balance

For Twitter users trying to build audience, personal authenticity and audience expectations must be balanced. To appeal to broad audiences, some popular Twitter users maintained that they had to continually monitor and meet the expectations of their followers. However, given context collapse, their followers had different preferences for revealing personal information versus focusing on informative topics. Our respondents described an ongoing loop of impression management as they altered this mix based on audience feedback.

Soraya Darabi, the social media strategist for the *New York Times*, said, 'I'm constantly aware of my followers.' She uses tools like Twittersheep (http://www.twitter-sheep.com), developed by her company's research and development staff, to track what her 472,000 followers are interested in. Soraya knows her audience is interested in 'media and marketing', so she focuses on those topics. At the same time, she tries to interject her own personality and passions—like music—to retain an authentic voice. Soraya said:

> Say you're an author, a book aficionado. Most [of your followers] have tagged music as a passion. You might want to throw them a bone about your favorite song. There are a lot of Venn diagram overlaps in this community. It's to your advantage to be as much as part of a community as possible which means engaging with people's interests.

Brandon Mendelson, an activist with more than 700,000 users, defined his audience as 'my "tribe," people who are interested in leading change in their organizations or day-to-day life by using emerging technology and people interested in helping others through social networks'. Brandon primarily focuses on subjects that appeal to this tribe, but he agreed with Soraya that a mix of personal and professional is necessary for active engagement on Twitter:

> Occasionally I'll get a person not happy about how often I tweet, which is quite a bit, and if I tweet about something personal about my college plans. I always tell them I tweet about what I want to tweet and that social networks are a personal platform. By not sharing personal information I'm not building a strong relationship with my audience.

Other respondents' comments echoed this view:

> Authenticity Rule 1: Include personal w/promotional. "Bags under my eyes from from staying up 'til 4 accepting friend requests."
> to me, authenticity means being human; tweets include mix of ups, down, personal, professional. v.little robot or corporate speak mix of work and social is interesting; agree it creates authenticity, but some find it annoying/distracting

'Corporate-speak' or 'work' topics were seen as less authentic than personal, 'human' revelations. However, the intimacy of these revelations is limited. Note that both Soraya and Brandon's examples of personal topics are relatively innocuous. Their decisions to reveal personal information are strategic, and often framed as a way to reinforce relationships with followers. Soraya noted:

> I don't put romantic or deeply personal information on Twitter. I do say when I'm spending time with X who they are, but typically that person is in new media and it may look good for professional purposes to say I'm having lunch or dinner with X. It also serves as a call to action to the newsroom that person X is in town.

These exemplify highly self-conscious identity presentations that assume a primarily professional context. Revealing personal information is seen as a marker of authenticity, but is strategically managed and limited. Similarly, several respondents mentioned that concealing personal information was a way to avoid alienating followers, deliberately avoiding topics that their followers might not agree with:

> I try very hard not to Tweet hate speech, anything divisive, try to send messages that will bring people together

politics and religion can be a little dangerous, 'cus you never know which of your followers you might offend.

Keeping balance is tricky; both Soraya and Brandon mentioned criticism from real-life friends. Soraya has a few close friends whom she asks to critique her Twitter stream to ensure she is striking a good balance: 'I run blind checks—do I look too much like a marketer? Am I tweeting too personally?' The mix of personal and informative tweets from users like Soraya and Brandon allowed them to maintain multiple audiences that included both personal friends and professional contacts. Rather than appealing si-multaneously to multiple audiences, each mixed tweets with different target audiences to maintain their broad appeal. This technique resembles the polysemic and coded communicative strategies of image-management experts.

Micro-celebrity, conceptualized as a learned practice supported by the infrastruc-ture of social media, can create tension. Twitter's directed friendship model replaces 'friends' with 'followers' and prominently displays the number of followers on each person's Twitter page, creating a quantifiable metric for social status. The ability to stra-tegically appeal to broad audiences and retain the attention of others is publicly valued through third-party services that rank people according to their number of followers. Micro-celebrity practices like interacting directly with followers, appealing to multiple audiences, creating an affable brand and sharing personal information are rewarded, and consequently encouraged, in Twitter culture. The ability to attract and command attention becomes a status symbol.

At the same time, micro-celebrity practice can be seen as *inauthentic*. When asked to describe 'authenticity' on Twitter, respondents placed it in direct opposition to strategic self-promotion:

> When I present the concept of authenticity I usually mean no marketing speak, don't pretend you know everything. Be yourself.
> High honesty about what you're here for. Don't pretend to be my friend if you're here for promotion. (Promo is fine. Lying isn't.)
> Once I feel they've crossed the threshold of caring more about status/follow-count or trendy topics than their followers.

This view of micro-celebrity practice assumes an intrinsic conflict between self-promo-tion and the ability to connect with others on a deeply personal or intimate level. Some view strategic audience management as dishonest 'corporate-speak' or even 'phony, shameless promotion'. The encroachment of presumably profit- or status-driven 'public' techniques into 'private' social spaces is met with stiff resistance from people used to interactional norms that do not involve the commodification of social ties. We might ask if 'public' space is becoming synonymous with 'commercial', and if alternative mod-els of publicity and attention can thrive within the networked audience environment.

From the Broadcast to the Network

Twitter is an example of a technology with a *networked audience*. Media audiences are always imaginary, whether they exist in the writer's mind or as the target demographics for a sitcom. But while Fiske (1989) argues that the broadcast audience is a fiction meant to serve the needs of media institutions, the writer's audience services the *writer*. These two models of audience help contextualize the networked audience and its impact on online social behavior.

The Writer's Audience

Writers have long grappled with conceptions of audience because writers, unlike speakers, are separated from their audiences. In his essay 'The writer's audience is always a fiction', Walter Ong (1975) argued that writers imagine an audience appropriate to their topic and form, and use textual cues to write that audience into being. Writers write for and to this fictionalized audience, adapting to their imagined expectations. Lisa Ede and Andrea Lunsford (1984) distinguished further between the *audience addressed*—the actual readers of a piece of writing—and *audience invoked,* the audience constructed by the writer. Published writers are often told to tailor books to particular demographics; these 'future readers' are a fiction about the audience addressed.

The imagined audience of social media strongly resembles Ong's fictionalized audience. While Facebook or Twitter users don't know exactly who comprises their *audience addressed,* they have a mental picture of who they're writing or speaking to—the *audience invoked.* Much like writers, social media participants imagine an audience and tailor their online writing to match.

The Broadcast Audience

Today, when we think about 'audience', we imagine people watching movies in a theater, or at home watching television. This model of the audience has historically been viewed pejoratively in media studies—as an unidentifiable mass who passively consume (Livingstone, 2005: 24). It is also intrinsically *institutional*; broadcasters assume an audience that is anonymous, static, and geographically bounded (Drotner, 2005). One-to-many communication implies a single broadcaster distributing content through a complex entertainment structure that reaches audiences who cannot respond back.

The broadcast model has been complicated through studies of active audiences and through the fragmentation and dispersal of mass audience. Active audience theory maintains that the meaning of a media text is negotiated by the audience; rather than consuming blindly, audiences use interpretive lenses and bring individual experiences to bear when making meaning from media (e.g. Fiske, 1989; Radway, 1984). With the advent of cable television in the 1970s, the VCR in the 1980s, the DVD player in the 1990s, and the ubiquity of home broadband and video games, audiences have splintered. The top-rated shows on television are viewed by a fraction of the audience that watched 1960s network television; niche networks and targeted media have proliferated (Turner, 2009).

As a result, the idea of the 'audience' as a stable entity that congregates around a media object has been displaced with the 'interpretive community', 'fandom', and 'participatory culture', concepts that assume small, active, and highly engaged groups of people who don't just consume content but produce their own as well (Baym, 1999; Jenkins, 2005). In decentralized communication networks like mobile phones and email, 'audience' describes how a communicative medium mediates a relationship between content producers and receivers (Drotner, 2005: 196), requiring a more interpersonal and flexible model. New media has changed the broadcast model of the audience, decentralizing media production and distribution (Benkler, 2006). The network changes it further.

The Networked Audience

The networked audience combines elements of the writer's audience and the broadcast audience. It consists of real and potential viewers for digital content that exist within a larger social graph. These viewers are connected not only to the user, but to each other, creating an active, communicative network; connections between individuals differ in strength and meaning (Haythornthwaite, 2002; Boase et al., 2006). Just as the broadcast audience flattened separated demographic groups into a mass audience, the networked audience combines a person's social connections, revealing the fiction of discrete face-to-face audiences. While the broadcast audience is a faceless mass, the networked audience is unidentified but contains familiar faces; it is both potentially public and personal. Like the broadcast audience, the networked audience includes random, unknown individuals, but, unlike the broadcast audience, it has a presumption of personal authenticity and connection. Social media participants are far more concerned with parents or employers viewing their Twitter stream than a complete stranger.

In contrast to the imagined broadcast audience, which consumes institutionally-created content with limited possibilities for feedback, the networked audience has a clear way to communicate with the speaker through the network. This opportunity for communication influences how speakers respond and what content they create in the future. Audience members take turns creating and producing content, and in this 'many-to-many' model the network constantly centers on who is talking, responding, or replying. Social media environments become a place where person-to-person conversations take place around user-generated content amidst potentially large audiences.

The networked audience contains many different social relationships to be navigated, so users acknowledge concurrent multiple audiences. Just as writers fictionalize the audience within the text in their *audience addressed,* Twitter users speak directly to their imagined audience. For instance, some ask their followers questions that assume a particular collective knowledge. They target tweets to specific audience members, and conceal or reveal information based on who they imagine to be listening. Some construct a sophisticated model of who may be reading their tweets based on linguistic, cultural, and identity markers in their Twitter stream. Managing the networked audience requires monitoring and responding to feedback, watching what others are doing on the network, and interpreting followers' interests. The network is therefore a collaborator in the identity and content presented by the speaker, and the imagined audience becomes visible when it influences the information Twitter users choose to broadcast.

Networked media brings the changes Meyrowitz described to interpersonal interactions. In sites like Twitter and Facebook, social contexts we used to imagine as separate co-exist as parts of the network. Individuals learn how to manage tensions between public and private, insider and outsider, and frontstage and backstage performances. They learn how practices of micro-celebrity can be used to maintain audience interest. But Twitter makes some intrinsic conflicts visible. On the one hand, Twitter is seen as an authentic space for personal interaction. On the other, social norms against 'oversharing' and privacy concerns mean that information deemed too personal may be removed from potential interactions. Similarly, the desire to have 'fans' or a 'personal brand' conflicts with the desire for pure self-expression and intimate connections with others. In combining public-facing and interpersonal interaction, the networked audience creates new opportunities for connection, as well as new tensions and conflicts.

Section 2

Where We Live

The Digital Home

A New Locus of Social Science Research

By Anne Holohan, Jeanette Chin, Vic Callaghan, and Peter Mühlau

Introduction

Digital Lifestyles and Homes: A Multidisciplinary Vision

The vision for a digital lifestyle refers to the extensive integration of computing and communication technologies into our everyday lives, to such an extent that people and technology form symbiotic relationships, supporting an ever richer, more engaging, and deeply connected set of experiences. The notion of lifestyle implies the idea of change; that our "digital lifestyles" (our behavior, expectations, etc.) change incrementally with technological innovations and acceptance. Also, the consumption of digital technologies, and to some extent, lifestyle, is about choice (either unconscious or conscious). Thus, the vision for a digital lifestyle is highly multidisciplinary in nature embracing social and physical sciences in equal measure. Lifestyles are rooted in real-life environments such as homes, offices, transport, and cities where people live. It is possible to use real environments that ordinary people inhabit as test beds for new technology, a concept that has been dubbed a "Living Laboratory" by the EU who have established a network of facilities labeled the European Network of Living Laboratories (http://www.openlivinglabs.eu/). The iSpace used in this paper is an example of a Living Lab and is part of this European network.

An example of a digital lifestyle environment is the digital home in which most electronic appliances and systems feature a network connection. The arrival of Internet has enabled all manner of home appliances and devices to be connected to networks making it possible to adjust heating, open shutters, check that lights are off, start the washing machine, set the alarm, show who is at the door, check that a baby is asleep, and feed the pets even when away from home. Moreover, by connecting information and media sources to networks, it becomes possible to create new forms of entertainment, work or play by, for example, delivering films on demand or teleconferencing among friends.

Delivering this vision is not easy, and is the subject of ongoing research with many technical issues being considered such as end-user programming, security, reliability, and maintenance (Callaghan et al., 2007). Thus, there are also many pitfalls that can lead to poor acceptability and, consequently, living digitally poses a number of challenges to designers and users on a variety of technological, social, political, and cultural levels that we will discuss as part of this chapter (Johnson, Callaghan, & Gardner, 2008). In order to investigate the possibilities for such digital homes, a handful of test beds have been built around the world to explore future visions where there is the progressive "instrumentation" of social environments through, for example, mobile and ubiquitous computing devices. In this paper, we draw on work from one such digital home test bed: the iSpace based at the University of Essex. The iSpace is a two-bedroom apartment built from the ground up to allow experimentation with new network-based home appliances and systems.

In this chapter, we propose that such digital homes offer an opportunity to provide useful social research data in a number of ways. Most basically, we can gather behavioral data that is produced as a by-product of a digital home. However, as the technology allows for innovative use of this data by households themselves, a rich seam of data on collective behavior and decision making for social research can be made available. For example, a family might govern the operation of their home appliance (manage) to conserve energy that would benefit them (lower energy bills) and the community (less greenhouse emissions). The smart technology (novel forms of sensors, agents, human-computer interaction (HCI), and networks within the home), under the management of the home occupant, collects usage data that is transmissible by Internet. From a single household level, and with the agreement of the occupants, the data can be transmitted to a variety of agencies (including for social research purposes), and Internet forums can be used to engage with other households to produce a collective for managing resources that is a blend of community and organization.

For social science, the greatest potential interest lies in the theoretical significance of this innovative use of technologies, but the greatest challenges are methodological as we will discuss later in the paper. The implications for social research go well beyond examining usage and behavior patterns of particular appliances. The digital home potentially offers one insight into a revolutionary way of organizing society as already argued by some in social science. In the 1960s and early 1970s, sociologists suggested that industrial society would give way to the "information society" with consequences in all fields of human endeavor (Bell, 1973). Later, Manuel Castells (1996) argued that the world is entering an "information age" in which digital information technology "provides the material basis" for the "pervasive expansion" of what he calls "the networking form of organization" in every realm of social structure (p. 468). He predicts new forms of identity and inequality, submerging power in decentered flows, and establishing new forms of social organization. More recently, Hardt and Negri in *Empire* (2000) have considered the consequences of the computer revolution and the widespread adoption and development of ubiquitous, pervasive, and ambient computing, which they argue will lead to a totally new postindustrial, informatization mode of production. For example, they argue that wealth creation will move from the manipulation of physical resources (e.g., mining coal from the ground) to information resources (e.g., mining knowledge from large collections of computer-based information servers). The effects of this, in terms of population behavior, were also considered by Clarke, who considered various scenarios and the consequences for the population as a whole (Clarke et al., 2007).

This chapter will first outline the emerging technologies that are refining existing research methods and enabling new forms of researching social behavior. Given the reach of these new technologies, there are clearly ethical issues relating to privacy, confidentiality, and access that should be addressed and that we will discuss below. We conclude by discussing the implications and limitations of the digital home for social and market research.

The Smart Home and the Digital Population Observatory Model (as a Concept)

There have been several research projects concerned with designing systems for realizing digital homes. By way of a few examples, Georgia Tech's "Aware Home" (Abowd & Mynatt, 2005) and Microsoft's EasyLiving project (Brumitt, Meyers, Krumm, Kern, & Shafer, 2000) that have investigated context aware systems (systems that present users with functions and options that change according to the users context). The Adaptive House project in Colorado (Mozer, 2005), The University of Texas Mav Home project, (Cook, Huber, Gopalratnam, & Youngblood, 2003), and the University of Essex iSpace (Hagras, Doctor, Lopez, & Caliaghan, 2007; Doctor, Hagras, & Callaghan, 2005) have investigated a variety of artificial intelligence techniques to model user behavior and preemptively controlled the environment to meet the users' needs creating the so-called smart home.

A key feature of a smart home is that networked devices can be made to coordinate their actions to produce meta-services formed from communities of coordinating services, functions, and appliances. Thus, the home of the future will be a deceptively complex place containing tens or even hundreds of pervasive network-based services, some provided by physical appliances within the home, others by external service providers. Services could range from simple video entertainment streams to complex home care or energy conservation packages.

There are many visions for digital homes that speculate on how network services might change the nature of consumer products and peoples' lifestyles. The general expectation is that domestic services would be designed, packaged, and marketed by commercial companies. However, networked technology opens up the possibility to develop alternative models. For example, from a customer's perspective, it may be possible for end users (homeowners) to compose the functionality of digital homes based on aggregating coordinating sets of networked services. Such descriptions of composite services and their behaviors would form "virtual appliances" that could move with people as they migrate across differing environments (e.g., via the network, or contained in mobile phones) instantiating these functions wherever possible. For example, a homeowner could create a "DVD *Ambience*" virtual appliance that sets the ambient light level to a comfortable level whenever someone watches a DVD (by controlling the window blinds and the room's artificial lighting) or a "*Home Guard*" virtual appliance that checks and secures the home locks at bedtime. Naturally, there is a basic set of common needs that people have such as telephones, TVs, heating, etc., and these would form default "virtual appliances" in all homes. However, other, more novel, virtual appliances (composite service descriptions) created by lay people could even be traded between people as "innovations."

Further, this notion addresses the need in some people to be creative. For example, many people like to choose their own home furniture and wall coverings, personalizing their environments. The technology described in this paper enables these concepts to be applied to people "decorating" their own electronic spaces. In addition, this paradigm increases the control people have over the technology in their life, a requirement that numerous social research projects have reported as being essential to the acceptance of technologies in the home (Kook, 2003; Mäyrä, 2006).

The iSpace

The iSpace is a test bed for future digital or smart homes based at the Essex University campus. It takes the form of a two-bedroom apartment, containing the usual rooms for activities such as sleeping, working, eating, washing, and entertaining. It comprises numerous regular but networked appliances such as telephones, media players, plasma screens, washing machines, refrigerators, lights, heating systems, etc., together with newer technologies such as speech and vision interaction (for commanding or configuring the environment), which interact with the user and sensors (e.g., detecting status of locks, food stocks, etc.) and tags (for tracking objects).

The iSpace (the home) is situated inside a larger community framework (the iCampus), which includes a wider instrumented environment (bars, shops) and wearables. The iSpace or home is an example of the wider concept that embraces other environments such as offices, shops, etc., showing that methods can be extrapolated to the wider networked environment domain.

In the next section, we provide a brief overview of iSpace technology that enables people to create micro services and government within the home, which we argue, can be extrapolated to inform macro government models and policies.

The Technical Methodology

A core tenet of the hypothesis we are arguing is that there are useful parallels between people forming policies to manage, for example, energy usage in their home, and policies being administered by local or central government. For this strategy to be effective the technology needs to maximize home owners' choice and control.

A number of significant studies have investigated digital home requirements. For instance, the Samsung Corporation, in cooperation with the American Institutes for Research (AIR), conducted a study aimed at identifying smart-home requirements by interviewing and monitoring people in South Korea and the United States (Chung et al 2003). One particularly important requirement discovered by the Samsung and AIR study was the need for people to be able to customize their home, a finding supported by many other studies (Mäyrä et al., 2006). There are many visions for digital homes that speculate on how users might be empowered to customize the electronic functionalities of their own home. An example of such a concept is that of MAps (meta-appliances and applications). MAps are "soft objects" that provide a means to aggregate elemental network services together to create virtualized forms of regular (e.g., TVs, air-conditioning, etc.) and novel (user created) appliances (Chin, Callaghan, & Clarke, 2009).

In terms of social science, MAps are uniquely useful as they form documents that described a person's real preferences with respect to some application or topic. For example, for energy policies, they would describe how a person set their heating in the home office, or transport for a variety of differing contexts such as lone use, family use, or business. Thus, they are a rich and accurate source of data, albeit at a somewhat low and detailed level. However, they are amenable to higher-level aggregation over hundreds or thousands of people, potentially providing more meaningful higher-level statistics.

A key consideration to the design of the technology is how MAps are created (Callaghan et al., 2005). Chin has researched this issue and makes a strong case for nontechnical end-user methods that she sees as a key requirement to empowering the principle stakeholder of the digital home to engage with the technology (Chin et al., 2009). She points at two solutions, one based on highly automated autonomous agents that monitor people's behavior, using this historical information to build models that aim to configure the technology to meet the person's future needs. She argues that such approaches have drawbacks such as a reluctance of many people to allow their personal home spaces (and them) to be monitored by network-connected technology in this way and, second, that trying to second guess needs based on past experience will inevitably have annoying failures (as not all future needs will be described by past actions) and it does not allow creative thoughts for novel MAps that might exist in people's mind to be efficiently extracted. For those reasons, she has argued for and produced a prototype system referred to as PiP (Pervasive interactive Programming) that users can use to create MAps by demonstrating, in explicit teaching sessions, the behavior they require (for full technical details and needs see Chin et al., 2009). Being able to gather data on people's needs, by directly monitoring what they create (as against what people say they may like based on more abstract judgments), has the potential to provide more reliable, detailed, and timely information (Rowley & Wilson, 1975). In particular, our approach builds on "gaming" and the "priority-evaluator approach" to social research, which uses analogues of the real world (frequently in the form of games) in which people are given the freedom to configure resources as a means to obtain more reliable social research data (Rowley & Wilson, 1975). In the approach being advocated in this paper, we retain the idea of monitoring how people configure resources as a means to elicit peoples' choices and desires but dispense with the use of analogies in favor of using new generations of configurable technology, in particular digital homes and PiP, which directly involves the stakeholders being studied. Incidentally, Rowley's work is also relevant to this paper as he is also credited with the earliest use of bespoke portable microcomputer technology to augment social and market research, concepts that further motivate this research (Rowley, Barker, & Callaghan, 1986).

Concerning the data yielded by such end-user customization systems, because people are able to create and design the functionality of their own virtual appliances or environments, their decisions and design actions represent valuable information in terms of understanding peoples' needs. This is especially so because the networked nature of the system makes collection of this data relatively straightforward for both individuals and aggregations of large numbers of people. However, rather than the data being conventional questionnaire-style answers, it is more complex as it contains details of assemblies (MAps) of functions, together with their usage. Inferring what the MAps do and why the user created them require meta-data tagging by the user at the time of creation. Usage statistics (when certain functions are used and for how long) can be automatically collected by the system. MAps and end-user programming are just one approach, albeit an important approach, to governing digital homes.

In summary, with respect to social science research the two methodologies described above (end-user programming and autonomous agents) provide a source of data on peoples' behavior and preferences. The agent-based approach represents more passive data gathered from monitoring peoples' behavior with little active input from the person being monitored. In contrast, end-user programming (and its PiP implementation) involves active participation of the home occupant in making decisions and forming policies. As such, we argue that the data gained from end-user programming is more linked to conscious decisions and policy making and thus more useful for extrapolating higher-level policy information. Decision making and its relationship to our hypothesis is discussed further in the following section.

The Model of Governance and Its Implications for Social Research

Emerging technologies such as the digital home have the power to disrupt and transform existing social structures and social practices. The "digital home," through networking appliances inside the home to each other and beyond the household via the Internet, offers the opportunity to generate usage data and allow people to manage and share that data via the Internet in a way that will have a transformative impact on consumption and decision-making patterns. The most innovative aspect of digital homes is arguably their ability to empower households to manage their own usage data and to deliberate with other households over whether and how to share this data with outside stakeholders, such as energy companies or government agencies. This disaggregates, and has the power to allow innovative reconstitution of, existing decision-making practices and relationships between providers of resources and consumers, and governments and citizens. Existing resource management and e-deliberation solutions can provide a means through which users can form and federate groups that can then electronically manage, reflect, and debate the best use of ambient intelligence resources in servicing their own needs and that of the wider community. By rendering such activities explicit, but by also providing user friendly Web tools to browse and manipulate the related models, communities can be enabled to decide clearly how to collect, monitor, process, and distribute sensor data.

Managing a home is in many ways analogous to the process of government. There are finite resources with much competition for them. Opinions and information need to be gathered, deliberated on, rules decided on within households (effective formation of policies), and actions taken (cf. a micro-government) (Callaghan, Clarke, & Chin, 2008). The household can engage in deliberation on how best to manage the household resources and then go online, if they wish, to engage in deliberation at three levels to exchange information and views on how to manage their data, and how that data is used: within the household between individuals; with other households in a selected group, specifically the street/neighborhood, municipal level; and between citizens and interested parties and government. This approach enables householders/ citizens to manage their own bounded domain—their own data—and if they wish to respond as a household or as a collective to specific policy and engage in deliberation on this policy with the government and other citizens. The government in turn could modify its policies in response to the findings and views of the households. These "micro-government" choices provide an opportunity for households to negotiate with commercial companies or state providers of resources, making the relationship much less passive and hierarchical than it is now. Taking the example of management of energy, existing technology can provide the means to monitor the power consumption of key appliances such as heating/cooling, lighting, hot water, washing machines, or TVs. Monitoring the individual power appliances is the minimal level of granularity needed to gain knowledge of personal behavior. This data from a digital home can feed policy formation as the data recorded can be transmitted to relevant parties/actors, for example, local municipal authority or central government. One can imagine a scenario where households, through negotiation, not only reduce their energy bills through the appliance specific data they are collecting but also could collectively share data on energy use with energy providers in exchange for lower fuel/energy charges. This would have implications for energy policy, including strategies to reduce the consumption of energy.

The emerging technologies used in "digital homes" offer opportunities to interrogate classical and contemporary social theories about the nature of social order. Contrary to some arguments in the Marxist tradition (Braverman, 1974; Schiller, 1976), rather than the information age leading to the increased control of capital through surveillance and deskilling and the further disempowering of labor, our model is part of a potential scenario where traditional power holders (capital, state) have some of their power decentralized or redistributed (Negri, 1988)—in this case, drawn down to households. IF (our emphasis) the lay user/worker/citizen is part of the process from the moment of inception, including easily customizable sensors, and IF there is use of the Internet forums for coordinating with other households, there is a significant new role for households and collectivities of households in the economy and polity. Households will have the information or data that the companies or state will require for efficiencies and this will provide bargaining power to previously passive consumers. The low level of cost for overcoming the hitherto material barriers for production of information (affordable sensors, access to Internet) means we have a scenario where anyone who is computer literate at a basic level will be able to own and manage their own information. Our model is aware of and tries to address this issue of access by including the means for customizing the sensors that does not require any technical knowledge.

With the ownership of data shifting to the individual, particularly the household level, and the necessity of negotiation for companies and bureaucracies who wish to gain access to the level of detail gathered, the structure and practice of institutions is impacted. Institutions, from household to companies to government, have, since the industrial revolution, been predominantly characterized by hierarchical structure and top-down authority relationships. The information revolution and postmodern society have provided the tools and consequences to challenge this in the family (Stacey, 1996), organizations, (Powell, 1990) and government (Keck & Sikkink, 1998; Slaughter, 2004). The rise of the "network society" (Castells, 1996, 2000) has changed the rules so that information flows now determine structure, boundaries are much less important (or indeed even possible), and status is determined much more by expertise than formal position (Holohan, 2005). Knowledge has never been more democratically distributed or available regardless of formal status. Politically, this offers a potential citizen-polity relationship that goes well beyond the franchise, and allows for deliberation and input into policy decisions that is unprecedented (Habermas, 1991; Calhoun, 1993).

The "networked information economy" (Benkler, 2006) is challenging the old "industrial information economy." This is producing a society where decentralized individual action, specifically new cooperative action carried out through radically distributed, non-market mechanisms that do not depend on proprietary strategies, is becoming increasingly important in the fields of communication, information, and culture (Lessig, 1999). For instance, the music industry has moved from dominance by huge music companies to a much more diffuse and democratized model of music publishing and performing. The material barriers are largely gone (due to technologies, principally the Internet) and non-market, nonproprietary motivations, and organizational forms are increasingly common, resulting in effective, large-scale cooperative efforts such as the

Open Source software movement. Eric von Hippels idea (1988, 2005) of "user-driven innovation" has begun to expand that focus to thinking about how individual need and creativity can drive innovation at the individual level and its diffusion through networks of like-minded individuals. It is this feasibility of producing information through social (cooperative and coordinated individual action) rather than market or proprietary relations that is at the heart of our proposed model of scalable governance of households using data that is produced by sensors and shareable with others through the Internet. So for instance, householders can choose to share information on energy consumption not on demand from companies or government nor in a market place but through collective deliberation and deployment of data for specific purposes, such as assisting energy efficiencies to address challenges from global warming. The collective deliberation could be at a street level or neighborhood level or municipality or city ward level. While there are threats to privacy (discussed below), there are also benefits. For example, by participating in energy usage monitoring programs, ordinary citizens can both help the environment and save money by using less energy. In addition, as such information might help make the energy providers more efficient, these savings could be shared with the consumer in the form of a discount providing additional incentives for the citizen to participate.

Privacy and Ethical Implications

The social sciences, and indeed society, need to grapple with the danger of technological developments proceeding at a pace that would realize the fears around privacy and "Big Brother" scenarios. The more attention social science gives to emerging technologies such as the "digital home," the more such dangers can be highlighted and addressed. Re-interrogating Marx and Weber, we can explore whether the new technology is increasing elite control of both politics and production through enhanced surveillance (Davis Hirschl, & Stack, 1997).

Numerous reports have revealed that user acceptance of technology in digital homes has been shown to be linked to perceptions of privacy, which in turn is linked to the degree of control the user has over the technology (Chung et al., 2003; Mäyrä et al., 2006). In terms of digital homes, these issues have been linked to the balance of autonomous agents versus manual management of the environment (Callaghan et al., 2008). A search of the Internet will quickly reveal that much has been written over the years about such dangers. This fear is increased as our control, and system transparency, is reduced. Thus if we are to live in digital homes, then questions such as "who has control?," "what is the extent of their control?," and "who has access to sensory data from our home and what use are they making of it?" are paramount. In terms of smart or digital homes, control comes down to the balance of technological autonomy versus user influence.

These concerns are graphically illustrated in the two-dimensional graph shown in Figure 28.2: the "3C framework" (Callaghan et al., 2008). To capture the balance of automation, there is an autonomy-axis that depicts the possibilities for configuration from manual (end user) to automatic (agent based). In terms of sociology, reactions to technology vary from love to fear, which are illustrated in an "attitude" axis that shows user reaction (philia versus phobia) to the different possibilities. The quadrants show differing combinations of technology and attitude, identifying potentially significant positions within this space. A general assumption underpinning this model is the view that the less understanding of and control over their technological environment people have, the more resistant or fearful they will be of it (and vice versa). The model is not normative but depicts a conceptual space of possibilities drawn from experiences of our research. Thus, for example, people that are technophobic may react to automated environments (where they have little say in how they operate) by trying to sabotage the technology. For example, people may cover or disconnect sensors to fool the system or prevent it monitoring them. In a system where technophobic people have control over the system, they may simply program the system to disable it, or make it work in an unconventional way, as a means of expressing their dissatisfaction. This is in contrast to the way people with technophilia tendencies may react. For example, for systems in which autonomous agents exercise total control over the environment they may marvel at the sophisticated technology and bask in the comfort it affords, whereas in a system that they can exercise control they may delight is creating novel functionalities. Of course these are extreme attitudes and behaviors and the diagram allows for more continuous variations.

Thus, in a digital or smart home the principal tool in the armory of privacy protections is control, which in practical terms resides in the balance of agent versus user management. By control or management, we mean maximizing the user's ability to make choices on what information is gathered and when and how it is used (including choices to "autonomise" the collected data). At a community level, this would require and enable groups at various levels of granularity (from families to neighborhood, town, and even to nations) to e-debate/deliberate what data should be made available to whom and for what purposes. Cheap storage, distributed systems querying, and perhaps peer-to-peer (P2P)-based backups could mean that sensor data could be stored relatively locally (minimizing exposure to massive theft and allowing the levels of security and robustness to be tailored to the group concerned).

In addition to people's desire for privacy, another barrier to people allowing detailed sensing of their domestic lives is rooted in a lack of trust in the officials or corporations who collect, store, have access to, and use such sensitive information. The 25 million personal data records lost in the United Kingdom in 2007 demonstrates the importance of addressing this issue. Local groups need to be able to negotiate about access of their data by "outside" bodies, for example, governments and corporation, from a position of power. These bodies would need to be forced to argue for access to data on the basis of earned trust, transparent procedures, and well-reasoned appeals to the common good or appropriate incentives. Agreement for access could be provisional (these are ongoing feeds of data so long-term relationships are key) and linked to systems for auditing security and usage and also for dealing with potential conflicts that may arise, for example, if one accessing body wishes to forward data to another one (a big concern with a lot of data privacy—you may trust the local police but not want them passing details to the CIA).

While in this paper we are focusing on the benefits of digital home technology, and how it might improve social research, which in turn could connect people and government in a more mutually effective relationship, it is equally clear that without careful planning and regulation of digital home technology it would be possible to create a modern equivalent of Bentham s Panopticon (Bozovic, 1995) or "Big Brother" (Orwell, 1949), turning homes into "gold-fish bowls" where our every move is monitored by third parties.

Blurring of Public and Private Spaces

As indicated above, it is already documented that the use of information technologies can itself pose a threat to confidentiality. The data collected and any accompanying communication that is typed rather than spoken leaves a physical trace referred to as a "data trace" that can be archived or preserved (Duffy, 2002, p. 85). Data traces can result in breaches of confidentiality if unauthorized people have access to research data stored on a computer that is connected to the Internet. In addition, in our example, protocols and standards for confidentiality need to be in place not just in the "smart home" but also in all relevant agencies: energy companies, government departments, market research companies, and social science departments in universities or research

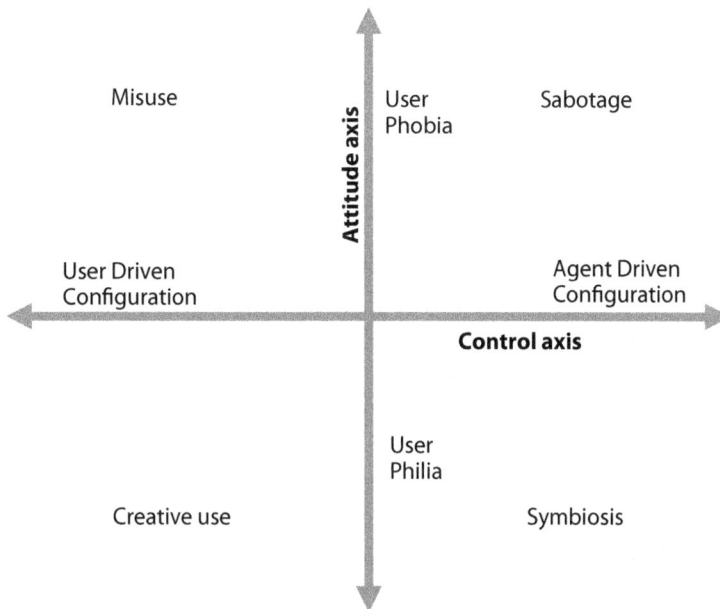

Figure 28.2– The 3C framework for digital homes

institutes. Such coordination represents a major challenge and can only really be addressed at the legislative level.

The difficulty in terms of the discussions via forums on the Internet, as to who and where and how to share the data, is the blurring of the boundary between public and private. Obviously, when people are in a public setting they can expect to be observed. Conversely, when they are in a private setting they do not. However, online this distinction is not clear (Barnes, 2004). Traditionally, there has been a clear divide between the public and private data with explicit consent required for inclusion of private data and measures taken to ensure that private data does not enter the public realm. But online flows of personal information are channeled through a public forum that poses no restrictions on audience access; anyone can view the discussion. Password-protected forums could be used, but this would restrict recruitment and participation in our proposed models; how else to entice neighbors over a relatively large geographical space, for instance to work together on sharing data, but to see the discussions and outcomes for oneself on the Internet? Traditionally, it is the discretion of the subject that determines the boundary between public and private. But messages exchanged online often create an illusion of privacy because contributors forget that other individuals can read those messages. Once individuals develop close Internet relationships, they can easily forget they are communicating in a public space (Barnes, 2004).

One practical solution is to ensure that while the public can access the deliberation forums, participation in the discussion requires membership. However, this does not deal with the problem of confidentiality for social science researchers, as participants using pseudonyms merely make the ethical problem one of dealing with multiple identities, as unless similar pseudonyms are substituted, online identities are not hidden if actual pseudonyms are written up in the research, as participants can be recognized by those familiar with their online pseudonyms (Beddows, 2008). This situation presents an as yet unsolved dilemma for Institutional Review Boards (IRBs), which exist to ensure the subjects of social research and participants in social research's rights are protected.

Situating "Digital Home" Research in Current Social Research

It is useful in the discussion of the potential of "digital/smart-house technology" to distinguish on the one side between *technology-specific* and *general research* and on the other between *market research* in a broad sense (including audience measurement and market research feeding into research and development) and *social research*. By technology-specific research, we refer to research that is generally concerned with the adoption, use, and adaptation of "digital home technology" or with the consequences of "digital home technology" for attitudes and behavior of the users and the social structure they are embedded in. With respect to the technology-specific research, the issues may overlap between market and social researchers although market researchers will be more interested in the adoption and use aspects while social researchers will be more tempted to examine the consequences of the technology. We have already addressed this, focusing on the implications for social research, in our discussion of the proposed model of disaggregated decision making and scalable governance.

General research is research that is interested in topics that are independent of the "digital home technology" and in a target population that is not the universe of actual or potential users of "digital home." Here, the perspectives of market researchers and social researchers diverge more strongly: Market researchers will be interested in the use of products and the market chances for new products, while social researchers will be primarily interested in the social organization of the domestic sphere. In the following, we focus on the potential and limitations for general research.

Most simply, the data collected in "digital homes" improves continuous measurement. The idea of continuous assessment is not new—for example diary surveys and audience measurement have been around for quite a while. However, the initiative to call in or complete the diary usually rests with the respondent. Technology has already been identified by researchers as improving continuous assessment and is giving much greater access into people's lives. For example, transaction data is already in use (credit or debit card use, video rental, etc.) and continuous measurement has been linked to the growth of portable Internet devices and mobile computing—mobile phones, Blackberries, etc. (Couper, 2003, p. 493). Traditional interviewer-administered surveys have very high costs for sampling, contacting, persuading, and interviewing, so interviewers have tended to maximize the interview and ask up to several hours of questions. Panel interviews, for the same high cost reasons, have tended to be months or years apart. However, there is existing evidence of the value of continuous measurement (Mundt, Bohn, King, 8c Hartley, 2002; Aaron, Mancl, Turner, Sawchuk, 8c Klein, 2004) and "digital home" technologies offer an opportunity to increase this value.

Collecting data in a "digital home" has potential for improving accuracy, as instead of surveying people's usage patterns in retrospect, where people report from memory, the hard data is much more accurate and complete. The pitfall of recall from memory is erased when you are working with real-time data; some obvious examples would come from research on diet, exercise, activities in the home, and mood. It also helps overcome observer fatigue: collecting data manually negatively influences the reliability of the data, apart from being very tedious, as the events of interest are rare and are interspersed with long periods of uninteresting activity (Philips Homelab http://www. research.philips.com/ technologies/projects/homelab/index.html).

Thus, the core of the primary data to be expected from "digital homes" appears to be process-generated usage data from linked and integrated equipment. It is useful to compare the potential of usage data with the very established industry of audience measurement (e.g., Napoli, 2003). The data generated by "digital home technology" is similar to what is known as "passive" audience measurement data. Traditionally, meters were attached to working television sets to record set-tuning data for the household. The use of these data was limited and was complemented by diaries of the individual household members and by survey-based information about the sociodemography and other characteristics of the household and its members. Diaries have been largely replaced by "active people meters" that share with diaries the dependence on the active cooperation of the household members resulting in data of doubtful reliability. More recently, the emphasis is on "passive people meters" allowing also the recording of media consumption outside the home and reflecting the use of multiple platforms, in particular of the Internet. The infrastructure for audience measurement is a "single-purpose technology" set up because the data generated are of utmost economic importance, in particular in the planning of advertisement campaigns and the price-setting for television commercials. To secure that the audience data permit reliable and precise population estimates, audience measurement requires panels that are representative for the population of television viewers. This is typically done by some form of probability sampling. Panels need to be regularly updated to continue to represent the population and validated by "external coincidental surveys."

What are the implications of this tale for the potential of digital homes in social research? Of course "digital homes" may be useful for the "passive" generation of multiple platform media consumption data at the household level, and individual meters should not be a problem. But are there other areas for which "digital homes" can produce valuable data? Energy consumption may be another area and an intelligent energy monitoring system may help the end user to use energy more efficiently. But the energy producer has far less interest than television companies in details about who consumes their product—the revenue of the utility company is independent of whether a male in the age group between 25 and 34 consumes their product or a woman in her 70s. However, unlike audience measurement systems, "digital homes" are not set up because of the need to collect data. To the degree that data about usage patterns are the by-product of their operation, setup costs are low and so are the marginal cost of generating (but not of processing) the data. Data generated by "digital home technology" may hence find interest from groups who would not be willing to invest massively in a data-generating infrastructure but see enough value in the data to cover the marginal costs and finance the processing of the data, for example, appliance manufacturers or home care providers. The integrated nature of the system, in addition, appears to permit one to examine interdependences in the usage of different components in the house. This may guide the analysis of substitution and complementary effects between different components and identify more general patterns of consumption ("lifestyles"). Moreover, new developments such as PiP are intended to allow users to translate mental concepts into designs of lifestyles or home functionalities they would like, but currently do not exist, with important implications for product research and development.

Limitations in this line of research are, first, that the natural unit of observation is the household and individuals have to be identified to permit data with individual household members as the unit of observation. Second, the research is confined to "home consumption," while there are obvious and important interdependences between "consumption at home" and "outside the home," as the trend in audience measurement toward "portable" meters recording mobile and public media consumption shows. Potentially, the core ideas of digital homes can be extrapolated to the wider networked environment domain easing this limitation. Third, for the foreseeable future, the weak point of "digital home"-generated data from the point of view of social research is the quality of the sample. High-quality samples are representative for the target population, and that is typically achieved by (1) a sampling frame that has high and unbiased coverage of the target population, (2) a sampling strategy that permits the generation of unbiased samples from this frame, for example, probability sampling, and (3) efforts to avoid low and biased cooperation and response rate.

Usually surveys are linked to individuals and thus there is the problem of how to differentiate individual from household in usage patterns. This has important implications for sample design and coverage. There is also not a stable sample of Internet users; unlike the real world, people, even if technically on an e-mail list, can choose to go "off-line." Participants can also mislead others about their physical location, identity, gender, or age, which means the researcher cannot effectively characterize his or her sample audience. Participation in Internet surveys and online deliberation, if sharing data from digital homes, will be affected by people working from different technological platforms; in particular, access to appropriate equipment and Internet technology and variation in bandwidth all impact who participates.

At the time being, inhabitants of digital homes are a small fraction of the overall population in terms of socioeconomic characteristics and age distribution but also in terms of attitudes and traits not representative for the general population. Because innovators typically differ from late adopters of technologies, current people living in a digital home are also different from the universe of smart-house digital home inhabitants (including future ones) and the former cannot be considered as a representative sample of the latter. "Digital home" owners are hence a sampling frame with low and highly selective coverage of the general population. There appears to be no incentive for data users to equip a representative sample of houses with digital homes technology in order to avail of representative data—that is, a crucial difference to audience measurement. Additional problems may be posed in securing the cooperation and willingness to share the data by people living in digital homes, as was discussed above. This may crucially depend on how "intruding" respondents perceive the "smart-house" technology if used for research and how sensitive respondents regard the data transferred. Such noncooperation is unlikely to be random. With ubiquitous deployment not yet established, as of now "digital home technology" might be valuable in providing reliable data of highly selective samples for narrowly circumscribed fields in marketing research.

Less clear is how large the potential of usage data is for social research in the broader sense. First, scope of the recorded usage appears to be excessively narrow to command much social-scientific interest. For example, in a research field like the domestic division of work, even individually attributed usage data cover only a narrow range of household activities. Activities that are not technologically mediated or supported are not covered; domestic work outside of the house is disregarded; and for many in-home activities of technologically mediated or supported activities, recorded usage is at best proxy measurement of the household task being completed. Second, in the case that a variable of interest is generated by the "smart-house technology," there are severe limitations of "what you can do with the data" *if it stands alone*. The scope for experimental manipulation is limited and quasi-experimental designs of a "natural experiments" type are the best one can hope for. Moreover, lacking random assignment as a mechanism to allocate people to treatment and control conditions, additional covariates have to be established to control for heterogeneity or allow matching of the groups. For observational studies, the problem of covariates is even more severe as (limited) causal inference requires measurement of the conditions expected to be the causal *agents* as well as antecedents and side-effect of this condition to eliminate spurious relationships.

All this would require that "rich" data are collected in addition to the process-generated data, using most likely "traditional" survey methods. For example, audience measurement data are not frequently used outside of their commercial applications and if so, then in highly aggregated form that permits one to link the data with other data sources (e.g., Hyland, Wakefield, Higbee, Szczypka, & Cummings, 2006; Frechette Roth, & Unver, 2007). Most studies on media consumption or media impact rely on self-collected data on media usage using either diaries (e.g., Couldry, Livingstone, & Markham, 2007) or self-reported usage (e.g., Chiricos et al., 1997), that is, data that are clearly less reliable than "passively" measured media consumption. A main reason is that social-scientific research is not particularly interested in purely descriptive studies of media consumption but either aims to untangle the processes generating different patterns of media consumption or to examine the effect of these patterns on behavior or attitudes. For both research interests, data generated by audience measurement are far too "thin" to permit meaningful social research and it is likely that the same holds for data generated by digital homes. Finally, most social research aims to be generalized to target populations and the quality standards are typically much higher in social research when compared with market research. Even if a highly salient dependent variable is recorded (or can be derived from data recorded by the "digital home technology") and relevant covariates are additionally measured, the fact that as of yet the inhabitants of "digital homes" are a highly selective group diminishes the value of the data for most social research.

The value added to qualitative research by "digital homes" is less obvious but nonetheless significant. As the researcher goes in with tremendous amounts of information already, it is much easier to orient the researcher theoretically and to focus follow-up questions and observation or just observation and in-depth interviews. Having accurate usage patterns is useful, but interpretative data is also needed to fully understand the context of the use of sensors. In addition, the online forums—and potential accompanying off-line interaction—would be subject to ethnographic analysis. Online ethnography refers to a number of related online research methods that adapt ethnography to the study of the communities and cultures created through computer-mediated social interaction, in particular observation, participant-observation, and interviews. Some have contested that ethnographic fieldwork can be meaningfully applied to computer-mediated interactions (e.g., Clifford, 1997) but it is increasingly becoming accepted as possible (Garcia, Standlee, Bechkoff, & Cui, 2009). In fact, the term netnography has gained currency within the field of consumer research to refer to ethnographic research conducted on the Internet (Kozinets, 2002, 2006a, 2006b). The limitations of ethnographies of online cultures draw from its more narrow focus on online communities, its inability to offer the full and rich detail of lived human experience, the need for researcher interpretive skill, and difficulty generalizing beyond the community under study. These and the challenge of scale and confidentiality face the researcher of householders participating in forums to collectively manage household data.

Most promising from a social research point of view, as indicated in the discussion of a new form of scalable household governance, is the potential of the technology to enable and record household decisions and their outcomes, such as rules, in a variety of ways. Research interested in understanding how decisions in households are formed and how contextual factors (e.g., income differences between the partners) affect the process and the outcomes would be keen to have access to data that would test models of household decision making (or joint decision making of a group of households). Model-testing research is also less dependent on sample representation, if the goal is to test theories about general mechanisms and is more concerned about the internal validity of the findings rather than the external validity (similar to laboratory experiments).

Conclusion and Future Directions

This paper has explored the hypothesis that emerging technology, in the form of the digital home, can provide a new way of exploring the implications of a digital lifestyle for consumption, policy, and social research purposes. We illustrated how these methods are implemented by describing a working prototype of future digital homes called the iSpace and a methodology that empowers lay people to customize the functionality of their digital homes called PiP. The ability of lay-users to customize products provides a powerful tool for market and social researchers to gain an insight to the needs and behavior of ordinary people. We also discussed how and when the data from a digital home might add value to existing social survey and marketing approaches. For social science, the greatest potential interest lies in its theoretical significance but the greatest challenges are methodological.

Most significantly for social theory, emerging technologies such as the digital home have the power to disrupt and transform existing social structures and social practices. The "digital home," through networking appliances inside the home to each other and beyond the household via the Internet, offers the opportunity to generate usage data and allows people to manage and share that data via the Internet in a way that will have a transformative impact on consumption and decision-making patterns. The most innovative aspect of digital homes is arguably their ability to empower households to manage their own usage data and to deliberate with other households over whether and how to share this data with outside stakeholders, such as energy companies or government agencies. This disaggregates, and has the power to allow innovative reconstitution of, existing decision-making practices and relationships between providers of resources and consumers, and governments and citizens. In effect, this innovation can transform institutions and impact the existing power structure in society.

Methodologically, for current social and market research, the value of the digital home lies in its ability to enhance existing methods or elements of methods such as continuous measurement and the generation of passive user data. Limitations in this line of research include the difficulties with the unit of observation currently being the household rather than the individual; it does not investigate the important interdependencies between consumption "at home" and "outside the home" and finally, the quality of the sample is currently irredeemably weak. Although the use of such data for social research is thus circumscribed by the hitherto experimental nature of digital homes and their as of yet limited availability to the general population, there is clear movement in industry and academia to address these issues, with exciting possibilities opening up for both social and market research as a result. In particular, from the research at iSpace discussed here, the possibility of accessing representative data of peoples customization of appliances and the patterns of usage combination is one that would be of great interest to market researchers if sampling was not as problematic as it is currently.

Both, the theoretical and methodological potential and challenge of the digital home need to grapple with the danger of technological developments proceeding at a pace that would realize the fears around privacy and "Big Brother" scenarios. The more attention social science gives to emerging technologies such as the "digital home," the more such dangers can be highlighted and addressed.

Acknowledgments

We are pleased to acknowledge the valuable contribution of Michael Gardner to this paper.

Neighborhoods in the Network Society

The e-Neighbors Study

By Keith N. Hampton

Netville

Netville was one example of a study of networks and the use of ICTs at the neighborhood level (Hampton 2001, 2003; Hampton & Wellman 2003). Netville was an experiment, an attempt to provide future levels of Internet connectivity and services to a typical middle-class suburban neighborhood, and to evaluate the impact of the technology on neighborhood social networks. Some aspects of the experiment were intentional: high-speed Internet access (10 mbps), online music services, online health services, and a variety of communication tools, such as a videophone, instant messaging, multimedia chatrooms, and a neighborhood email discussion list. Other aspects of the experiment were unintentional, such as the presence of an internal control group of residents who did not receive the technology but lived in the same neighborhood. Systematic observations of how the technologies were used were incorporated into the design of the experiment. A detailed network survey was conducted by presenting participants with a roster of adult residents who lived in the community. Participants were asked to identify those they recognized and how often they communicated. The formal network analysis was complemented by two years of ethnographic observations.

When compared with non-wired neighbors, those who received access to Netville's technology were more involved with their neighbors: they recognized three times as many, talked to twice as many, visited with 50 percent more, and called them on the telephone four times as often. While those with the technology had more ties and more frequent interactions in-person and over the telephone, relatively weak, not strong intimate ties formed as a result of the services. The large number of weak neighborhood based ties was also found to have supported residents' ability to organize collectively when dealing with local issues and concerns. Of all the technology that residents were provided with, they most valued the neighborhood email discussion list, and felt that it was most effective in building local ties.

How generalizable are the findings from the Netville study? Other studies of ICTs in neighborhoods of a similar size have found that the technology was not adopted as experienced in Netville (Arnold et al. 2003). Based on what we know about neighborhood effects, would the same results be found in different types of neighborhoods? Were the findings of Netville an artifact of cross-sectional research, or would a longitudinal study produce similar results?

Methods

Neighborhoods

Data for the e-Neighbors project was collected through a series of three annual surveys. The survey was administered in 2002, 2003, and 2004 to the adult residents of four Boston area neighborhoods. The neighborhoods were selected to be socioeconomically homogeneous (middle-class) but to contrast in terms of housing type. The expectation was that housing choice would serve as an indicator of stage in the life-cycle and propensity and availability of community involvement (Michelson 1977).

Two of the four sites were located in the Boston suburb of Lexington. Located in the same US Census tract, both neighborhoods had low-density, single-family, detached homes built from the 1960s–70s. The neighborhoods were selected because of their geographic proximity to each other and because they each had identifiable neighborhood boundaries (boarded by forests, a lake, and major roadways). The first site consisted of 209 homes, the second 226 homes. Neither neighborhood had an existing home owners' association. An initial pre-survey investigation using census tract data confirmed suburban, middle- to upper-middle-class status: median household income US$94,000, high educational attainment (67 percent with a Bachelor's degree or higher), high home ownership (84 percent), low mobility (70 percent had not moved in the past five years), few households with one person living alone (23 percent), and 51 percent of family households had related children at home under the age of 18 (US Bureau of the Census, 2000 US Census).

The third site, a 23-story, 174-unit apartment building, was the product of 1960s urban renewal and was located on the site of Boston's former West End (Gans 1962). The apartment building had no formal or informal tenants' group. The median household income for the census tract containing the building was US$52,000, 72 percent of households had only one resident, 61 percent had a minimum of a Bachelor's degree, 72 percent lived alone, and only 11 percent of families had children under the age of 18. Compared with the suburban sites, this area was also significantly more mobile: only 32 percent of residents had not moved in the previous five years (US Bureau of the Census, 2000 US Census). One hundred percent of housing units in the area were occupied by renters.

The fourth site was located in the Boston suburb of Quincy; a 101-unit, medium-density, gated, multifamily condominium development built in the early 1980s. This development was one of the few neighborhoods in the Boston area that could be defined as gated. The development did not have barricades, but relied on close circuit television and security guards monitoring the entrance and open areas. As a condominium development, the neighborhood had a preexisting community association. Prior to contacting residents, a formal request to gain access and a formal presentation were made to the community manager and home owners' association. Based on census tract data the median household income of the gated community was US$59,000, 54 percent of residents had at least a Bachelor's degree, 40 percent of residents had not moved in the previous five years, 63 percent of residents owned their home, 62 percent of households were occupied by only one person, and only 8 percent of families had children under the age of 18 living at home.

Following the first survey, three of the four neighborhoods were given access to a series of experimental Internet services designed to facilitate access to local residents and local information. The fourth neighborhood, the second suburban site, served as a control group. Unlike community network studies that specifically set out to provide residential areas with computer equipment or Internet access, the goal here was to intervene as little as possible. To maintain as close as possible to the ideal of a natural research setting, participants were not given a computer, Internet access, or any training.

Technology

Residents of the three experimental neighborhoods that chose to receive the e-Neighbors services were provided with the following:

- A neighborhood email discussion list: Each neighborhood had its own email discussion list. Residents were sent instructions by email and postal mail on how to send messages to the list. Messages sent to the list were automatically redistributed to all subscribed addresses.
- A neighborhood website: As with the email list, each neighborhood had a unique website. Each neighborhood website was dynamically customized for each user (displaying his/her name and profile information) and neighborhood (displaying the neighborhood name and content provided by other participating members of their neighborhood). The website contained the following features:

  ◦ A user profile that included demographics, information on personal interests, and space for personal comments.
  ◦ A searchable neighborhood directory that included information from neighbors' profiles.
  ◦ A 'match maker' that matched participants based on common interests, hobbies, histories, etc.

- ◦ An instant messenger that identified which neighbors were connected to the Internet and available to chat.
- ◦ A community calendar.
- ◦ A forum to provide and comment on recommendations for local business and services. o A forum to list classified ads and items for sale.
- ◦ A 'community poll' that allowed participants to create multiple choice survey questions that were presented to other users of the neighborhood website.

With the exception of a reminder email and postcard sent every six months, which included the address for the neighborhood email list and the address/password for the neighborhood website, participants were not given any specialized training or prompting to use their neighborhood website or email list.

Use of e-Neighbors

Participants in the three experimental neighborhoods were provided with a neighborhood email list and website. To use the website participants had to sign in, create a personal profile, and return periodically to add content and see if other residents had added information. All participants were subscribed to the neighborhood email list and messages were delivered directly to their personal email accounts. It was anticipated that the integration of the neighborhood list with an existing means of communication, email, would reduce the cost of participation in comparison with the website. The email list would offer immediate visibility of participation that would overcome high thresholds of participation (Granovetter 1978) and facilitate the formation of a critical mass of active users (Markus 1987; Hampton 2003).

Tables 3 and 4 document the extent to which the e-Neighbors services were used in the experimental neighborhoods. In the apartment building there was almost no use of the neighborhood email list; in the first year one message was sent and no one replied. In the second year, no messages were sent. The gated community demonstrated more use of the neighborhood list, with eight residents sending a total of twenty-five messages in the first year, and two residents sending one message each in the second year. Comparatively, the suburban neighborhood demonstrated very high use of the neighborhood email list that increased over time: 42 residents sent a total of 115 messages in the first year, and 49 residents sent a total of 271 messages in the second year. The neighborhood websites were used by a very small number of residents in each neighborhood (Table 4). Activity on the website was limited to updating personal profiles and browsing the neighborhood directories. Unlike the neighborhood list, which experienced an increase over time in the suburban neighborhood, the websites experienced decreased use over time in all neighborhoods.

TABLE 3 Number of messages sent to neighborhood email list (number of senders in brackets).

	apartment	gated	suburb	control
year 1	1 (1)	25 (8)	115 (42)	n/a
year 2	0 (0)	2 (2)	271 (49)	n/a

TABLE 4 Number of visits to neighborhood website (number of visitors in brackets).

	apartment	gated	suburb	control
year 1	32 (5)	52 (10)	134 (11)	n/a
year 2	3 (2)	11 (2)	43 (6)	n/a

The one message sent to the apartment list was a question about construction on a roadway next to the building. Email to the gated list consisted of announcements and discussions of: local services, holiday greetings, local issues and concerns, meetings of the home owners' association, and the death of community residents. The two most common topics were the discussion of local services and announcements related to the death of neighborhood residents. The discussion of local issues included exchanges related to wild animals living near the community (who were occasionally accused of eating pets), mosquito control, and the change in location of a local voting station. The only evidence of collective action on the gated email list was a discussion to negotiate a group purchase of heating oil. Residents of the gated community particularly valued the email list in announcing the passing of neighborhood residents. As one female resident expressed, 'What a shock. Will you pass on the arrangements please? This shows the first real usefulness of this email list. I probably wouldn't have known otherwise.'

The suburban list was similar in content to the gated community, although there were no announcements related to the death of neighborhood residents (an indication of life-cycle differences between residents of the gated and suburban communities). The great majority of messages to the suburban list were requests and replies related to the recommendation of local services, including: electricians, plumbers, babysitters, home appliances, cellphone providers, Internet and cable providers, insurance agents and window washers. The following message was typical:

> Thanks for your previous home improvement recommendations. I get water in my sun porch, most likely because of poor/non-existent flashing between the roofs. Any recommendations for good roofers who do spot repairs like this?
>
> (Male Resident)

It was not until the end of the first year of the e-Neighbors experiment that residents of the suburban neighborhood first used the list to discuss local issues. At that time the town government conducted a special ballot for residents to vote on an override to a state law that limited property tax increases.

> To those of you who are interested in helping get the override passed. … Please contact Janice or Linda they are the women in our neighborhood working around the clock to make sure the correct information gets out regarding the override. They really need your help.
>
> (Female Resident)

While the majority of messages to the suburban neighborhood list were in favor of the 'yes' vote, there were exchanges between 'yes' and 'no' supporters. The exchange was conducted in a manner that could only be construed as polite, there was no flaming and no messages were sent to the list that could have easily been interpreted as uncivil or offensive. A drive through the town prior to the ballot revealed a significantly greater number of position signs on lawns in the experimental neighborhood in comparison with the control neighborhood, both in favor and opposed to the tax override. In addition to the 'yes' vs. 'no' exchange, an additional group of residents were opposed to the very idea of discussing politics on the neighborhood list:

> No More Political announcements please on the Neighborhood e-mail list. This great service is supposed to bring the neighborhood together but politics (and religion) often does the opposite.
>
> (Signed jointly by a husband and wife)

These messages in turn generated a series of messages in favor of using the list to discuss local political issues.

> Citizens shape their communities, and communities shape the nation, and so on and so forth. That's the real meaning of 'grassroots'. This email list is, to my mind, a forum for local citizens to discover common views and form grassroots groups to support actions or merely to explore common interests. We've all bemoaned how distant government has become from the real world—our world and our daily activities. Perhaps this is precisely the technology that can bring 'participation' back into the concept of democracy. How else can we make a difference? National and international issues, as well as local issues, ought to be fair game prior to elections. But not as a soapbox: this list needs to be respected as an ongoing conversation between many parties, not a bully pulpit for a vociferous few. Nor should exploration and connection be limited to politics. For example, speaking for myself, I've been involved professionally, for the last few years, with gender equity. Is there any one else out there who's interested in, or actively involved with gender equity? I'd be interested in knowing who you are; please drop me a line. I don't have Jim's flair with quotes, but I am reminded of the song from 'The King and I': 'getting to know you, getting to know all about you (la da de dah dah, la da de dah …)' Otherwise known as community-building, and it's a very civilized use for technology.
>
> (Female Resident)

In total fifty-three messages were exchanged over the eleven days immediately preceding and after the town ballot. No residents asked to be removed from the discussion list as a result of the discussion. In the year following the ballot, the content of email messages on the suburban list diversified to include both service recommendations and discussions of local issues, including a local bus service, local schools, town elections, voter participation and an additional tax override. Residents also used the service to organize local events, including an annual garage/yard sale and a number of small neighborhood gatherings.

Given the limited use of the intervention in the apartment building and gated community, there is no reason to expect that e-Neighbors had any significant or enduring impact on network size, closeness or communication between residents. Therefore, the remaining analysis of how neighborhood networks changed over time is limited to a direct comparison of the suburban experimental and control neighborhoods.

Closeness

Closeness is a measure of tie strength (Marsden & Campbell 1984). ...[8]

At the beginning of the study, residents of the experimental neighborhood had as many close ties as other suburban participants. Controlling for demographic characteristics, the average person felt close to 11.82 neighbors. However, in comparison with the control neighborhood, residents of the experimental suburb were close to a smaller proportion of their network (pre-intervention); they were close to 6.69 percent less of their network. Men felt close to a similar number of neighbors to women, but men's networks were more intensive; they were close to 7.12 percent more of their network. Those with children and those who had lived in the neighborhood longer were close to both a larger number and larger proportion of their network. Years of education was associated with fewer (−0.58/year) and a lower proportion (−1.54 percent/year) of close neighborhood ties.

There was no observed change over time in the number or proportion of close ties for those with or without the e-Neighbors services. Not only was there not a general trend toward more closeness with neighbors as a result of the experimental intervention, but prior to seeing the neighborhood roster, when asked how many new 'close' ties they had developed as a result of the technology, only one participant indicated that they had a new 'close' neighbor (Table 6). Only three participants in the first year and five in the second reported making at least one new 'friend'. The great majority of ties formed as a result of e-Neighbors were weak social ties.

Contact In-person

At the start of the study there was not significant variation between suburban neighborhoods in the number of relationships participants maintained through face-to-face contact. The average participant had 11.44 in-person contacts over the preceding thirty-day period. However, those who chose to enroll in the e-Neighbors services maintained face-to-face contact with a slightly smaller proportion (7.43 percent less) of their network (pre-intervention). Over time, both the proportion and number of network members contacted in-person decreased for those with e-Neighbors, regardless of the extent of their participation (sending email to the neighborhood list or not).

The average e-Neighbors participant lost 2.51 face-to-face encounters, 8.84 percent of the in-person contact with his/her network, in each year of the study.

It is unclear why residents who maintained face-to-face contact with a smaller proportion of their network self-selected to participate in e-Neighbors. However, it is likely that this self-selection was partially responsible for the trend of reduced in-person contact over time. The failure to find a difference between active and passive participants on the neighborhood email list suggests that those who engaged face-to-face with a smaller proportion of their network gravitated toward a technology that they felt would reinforce a need for even less face-to-face contact. e-Neighbors either passively allowed an exiting individual trend toward less face-to-face contact to continue, or it justified a further reduction in face-to-face contact amongst those who already had less contact with their network. Still, even though the trend was toward reduced in-person contact, prior to reviewing the neighborhood roster, eighteen participants in the first and eleven in the second year of the study reported meeting at least one new neighbor in-person for the first time as a result of e-Neighbors (see Table 6). Amongst those who met a new neighbor the trend was towards multiple new face-to-face encounters, over 50 percent reported meeting more than five new neighbors.

Contact by Telephone

Comparing control and experimental groups, there were no differences in telephone communication across neighborhoods at the start of the study or over time. Individuals tended to not change the number of neighbors with whom they communicated by telephone. There was a tendency for men to have fewer phone contacts than women (–0.74), and for those with children (1.02), and those with a very much longer tenancy (0.04/year) to have phone contact with more network members. Those who were married also tended to maintain phone contact with a slightly larger proportion of their network (2.33 percent). Prior to being presented with the neighborhood roster, when e-Neighbors users were asked if the technology they were provided with had changed telephone contact with neighbors, fourteen in the first year and nine in the second reported talking to a new neighbor on the phone as a result of using the service they were provided with (see Table 6).

Contact by Email

Prior to the e-Neighbors intervention there were very few emails exchanged between neighbors in either the control or experimental neighborhoods. Such email as was exchanged was intensive with a small number of neighbors. There was a statistically significant, although numerically small difference between the experimental and control suburban neighborhoods in the proportion of neighborhood ties emailed at the start of the study. Residents of the experimental neighborhood who chose to enroll in e-Neighbors emailed an average of 1.12 percent fewer of their local ties before the experimental intervention began. After the e-Neighbor intervention, only those who actively participated by sending email to the neighborhood email list also sent personal

TABLE 6 New ties recalled by those enrolled in e-nelghbors suburban neighborhood (N = 94).[a]

	new close	new friends	new in-person	new phone	new email
year 1	0	3	18	14	22
year 2	1	5	11	9	19
total	1	8	29	23	41

[a]Based on a series of five questions presented prior to the neighborhood roster that ask 'in the past year' how many neighbors have you 'met in person'/'talked to on the phone'/ 'emailed' for the first time as result of the e-Neighbors services? How many would you consider to be 'friends'? How many are 'close friends'?

emails to more of their network members (in addition to email to the neighborhood list). The increase in the number of ties emailed was modest, averaging 0.33 additional ties per year. Early Internet adopters in both the control and experimental neighborhoods experienced a similar small increase in each year of the study (the average Internet user with seven years of Internet experience gained 0.21 email ties per year in the study while someone who had only been online for two years gained an average of 0.06 email ties per year). Couples also had slightly more local email ties (0.24). Before seeing the neighborhood roster, when participants were asked how many new neighbors they had emailed as a result of the e-Neighbors services, 24 percent reported emailing at least one new neighbor in the first year of the study, and 21 percent reported at least one new email tie in the second year (see Table 6). Of those, 20 percent reported emailing five or more neighbors whom they previously did not know.

Discussion

This paper explores the contexts under which Internet technologies are capable of bridging electronic and parochial spaces to augment neighborhood social networks. It was hypothesized that the influence of an experimental intervention, consisting of a neighborhood email discussion list and website, would vary depending on neighborhood characteristics. In particular, neighborhoods with an existing propensity for local interaction, characterized by high residential stability, the presence of children, a strong sense of community and an interest in building neighborhood ties, would have a higher rate of adoption. In areas where the experimental services were widely adopted, it was hypothesized that the neighborhood email list would enable diverse neighborhood exchanges that would increase the size of neighborhood networks and the frequency of interactions on- and offline. In addition, it was hypothesized that Internet use in general was becoming increasingly embedded into everyday neighborhood interactions.

Of the three neighborhood settings tested through the e-Neighbors study, only one neighborhood widely adopted the interventions for use as local media: the suburban neighborhood. The intervention had the lowest levels of adoption in the apartment building; the neighborhood with the lowest level of residential stability, fewest co-habitating couples, lowest proportion of children, lowest rate of home ownership, little preexisting sense of community, and low sense of community obligation. Few in the apartment building visited the neighborhood website and only one message was ever sent to the neighborhood email list. This was despite the fact that the apartment building contained the youngest population of the four study neighborhoods, presumably the most technology savvy, and prior to the intervention the majority of residents—more than any of the other study neighborhoods—expressed a desire to have additional contact with their neighbors. The gated neighborhood experienced slightly higher levels of adoption than the apartment building, but at its peak adoption was still low and dissipated over time. In many ways the context of the gated community resembled the suburbs, but with a reduced focus on children. It was also the only study neighborhood with a preexisting neighborhood association and presumably existing channels of communication. Pre-intervention a significant number of gated residents were also not in favor of establishing additional contact with their neighbors. In both the apartment and gated community, neighborhood contextual characteristics overwhelmed any individual desire to use the technology locally.

The suburban neighborhood contained many of the characteristics identified by the neighborhood effects and social ecology literature as contributing to a greater propensity for local tie formation. The e-Neighbors services experienced high rates of adoption and, as anticipated, the neighborhood email list was used at higher levels than the neighborhood website. The email list was used to discuss local services, local politics, local issues, and collective action. Only those residents of the experimental suburb who actively participated in the neighborhood list experienced a change in the size of their networks, an average increase of four new ties in each year of the study. As hypothesized, new ties formed as a result of the e-Neighbors intervention were 'weak', not strong close ties. There was little support for the hypothesis that contact would lead to contact. Those who had less face-to-face contact in the experimental suburb before the intervention were more likely to adopt the technology, but adoption of e-Neighbors did not reverse the existing trend, the tendency toward less face-to-face continued and was possibly reinforced. There was no change as a result of the intervention in the number of neighbors contacted by telephone. However, those who actively participated, by sending messages to the neighborhood email list, began, although at a very limited rate, to email a larger number of neighborhood ties.

This study supports the hypothesis that Internet use is increasingly embedded into neighborhood networks, and in surprising ways. A significant relationship exists between time of Internet adoption and neighborhood relationships. Early Internet users had smaller networks at the start of this study, but their networks were more robust than others. Those who were amongst the earliest to adopt the Internet may have been more socially isolated than people in general, or a pattern may exist where initial adoption of the Internet corresponds to a drop in social capital that is ultimately mended over time. Unlike late and non-adopters, who experienced a decline in neighborhood network size over the two years participants were followed, early adopters experienced a reasonable level of growth in the size of their local network. In addition, the longer a person had been online the more neighbors he/she maintained contact with by email.

This suggests that Internet use does not privatize; it does not isolate people from the parochial realm of the neighborhood. Internet use over extended periods appears to be an antidote to privatism—it affords the formation of local social networks.

Section 3

Love and …

Intimacy

By C.J. Pascoe

In this chapter we explore teens' normative and nonnormative patterns of intimacy practices and new media. In doing so we sketch out the trajectories of historic and contemporary teen courtship rituals and the ways new media have become a part of these rituals, as well as highlight themes of monitoring, privacy, and vulnerability. Looking at these themes indicates that boundary work is a central part of navigating new media in intimate relationships. These intimacy practices also show how casual, friendship-driven use of new media might be a form of informal learning through which teens develop literacy by building relationships and communicating with their intimates.

Dating, New Media, and Youth

Given that teens have been the developers and shapers of contemporary youth dating culture (Trudell 1993), it makes sense that they would quickly put new media to use in the service of their romantic pursuits. While courtship norms and practices are less formal and more varied than they were in the early and mid-twentieth century, our research on teens' new media use shows that the rituals are no less elaborate or important than those of their historical counterparts. Dating and courtship, as enacted by contemporary American teens, is largely a twentieth-century development, as is the life stage of adolescence itself (Ben-Amos 1995). After the industrial revolution, when families declined in importance as economic units, romantic unions gradually superseded primarily economic ones as a social norm in the West. Middle and upper-class young people courted through processes heavily monitored by parents, families, and communities in which young men would "call" on young women in their homes (Bogle 2008). Dating, as we now recognize it, emerged out of working-class "calling" practices, in which young ladies lacked the domestic space to entertain young men in their homes and thus the couple would go out somewhere together, a practice referred to in early slang as a "date" (Bogle 2008). As the 1920s progressed, rebellious middle-class youth emulated these working-class rituals (Bogle 2008). These imitations, along with the movement of youth from workplaces to public schools, the development of school dances, and the independence afforded by the spread of automobile ownership, laid the groundwork for contemporary teen dating culture (Modell 1989). In the 1950s teen dating norms were formalized, became close to a universal custom in America, and were solidified by the practice of "going steady" (Bogle 2008; Modell 1989). Youth who "went steady" indicated to onlookers that they were unavailable by trading class rings, letter sweaters, ID bracelets, or by wearing matching sweater jackets—their answers, as one historian puts it, to the "wedding ring" (Bogle 2008, 17).

In the 1970s and 1980s, these types of formal dating and "going steady" practices declined as dating became "merely one form of social contact among many" (Modell 1989, 291). The decline in formality is reflected in contemporary teens' language about these types of relationships, which frequently lack a clear vocabulary to define relationship status or practices: "The terms *courtship* and even *dating* have given way to *hanging out* and *going out with someone*" (Miller and Benson 1999, 106). However, the decline in the formality and uniformity of dating practices does not mean that the centrality of romance to teenagers' lives has declined in salience. One study showed that the strongest emotion during puberty was "the specific feeling of being in love" (Miller and Benson 1999, 99), and developmental psychologists consider romantic relationships an essential feature of social development in adolescence (Connolly and Goldberg 1999). Contemporary relationships among teens tend to be "casual, intense and brief" (Brown 1999, 310). They are also, for all their emphasis on privacy and exclusivity, profoundly social (Brown 1999). In adolescence "peers provide opportunities to meet and interact with romantic partners, to initiate and recover from such relationships, and to learn from one's romantic experiences" (Collins and Sroufe 1999, 126). Especially in the early flirtatious stages, "romance is a *public* behavior that provides feedback from friends and age-mates on one's image among one's peers" (Brown 1999, 308). [1] Teens learn about dating, intimacy, and romance from their friends and social circles. Further, while we usually think of these intimacy practices as individual and private, teen romance and dating rituals take place, in many ways, publicly and collectively.

Dating and romance practices and themes, so central to contemporary American teen cultures, not surprisingly are a central part of teens' new media practices (Lenhart and Madden 2007; Oksman and Turtainen 2004). Using social media, contemporary teens continue to craft and reshape dating and romance norms and rituals that are now deeply tied to the development of new media literacies. Social media technologies have provided a more extensive private sphere in which youth can communicate primarily with age-clustered friends, acquaintances, and sometimes strangers outside the purview of their parents or other authority figures. These more private channels of communication have allowed an elaboration of teens' intimacy practices, especially in forming, maintaining, and ending romantic relationships. The familial negotiations over the spheres of privacy in which these practices take place will be elaborated upon in chapter 4.

In their intimacy practices youth use three primary technologies—mobile phones (though many do still use home phones), instant messaging (IM), and social network sites. Mobile phones provide youth a way to maintain private channels of communication, maintain continual contact, and also serve as a "leash" through which teens in a relationship keep "tabs on" one another. Teens use instant-messaging technologies to maintain frequent casual contact with their intimates. As described in chapter 2, social network site profiles are key venues for representations of intimacy, providing a variety of ways to signal the intensity of a given relationship both through textual and visual representations. While most of their online relationships map closely to their offline ones, these digital spaces give teens the ability to reach beyond institutional and geographic constraints to forge romantic relationships. All these technologies allow teens to have frequent and sometimes constant (if passive) contact with one another, something Ito and Okabe call "tele-cocooning in the full-time intimate community" (Ito and Okabe 2005a, 137). Many contemporary teens maintain multiple and constant lines of communication with their intimates over mobile phones, instant-message services, and social network sites, sharing a virtual space that is accessible only by those intimates.

Surprisingly, given its centrality to teen culture, very little has been written about teens' contemporary romance and courtship practices. Researchers have directed their studies of romantic relationships toward adults (Hartup 1999) and focused on teens' sexual practices (e.g., Ashcraft 2006; Martin 1996; Medrano 1994; Moran 2000; Strunin 1994; Trudell 1993). This research orientation likely reflects an American concern with teen sexuality as out of control and dangerous (Schalet 2000). In focusing on teens' intimacy, though not necessarily sexual, practices, we take a sociology-of-youth approach, following the categories and practices important to the teens we talk to, not allowing adult anxieties to guide our research. As a result we report little about teens' sexual experiences. Given the preoccupation with youth sexual practices, not to mention current popular concerns about sex predators and youth exposure to sexual content online, it seems odd to leave sex out of a chapter on intimacy practices. However, we simply did not hear a plethora of stories about sex in our interviews, as youth tended to discuss dating, crushes, romance, and heartbreak. This omission could be due to several factors. First, such intimate details might emerge in a second or third interview, which most researchers did not conduct. Second, we conducted these interviews under constraints imposed by our universities' institutional review boards, which heavily discouraged talking to youth in general and about issues of sex and sexuality in particular. Finally, it may be that intimacy practices were simply more salient to these youth than sexual ones.[2]

So even though romance is one of the focal points of youth popular culture, because of researchers' focus on sex, we know surprisingly little about teen romance, dating, and courtship practices, apart from scattered stories on historical dating practices

(Diamond, Savin-Williams, and Dube 1999). This chapter begins to remedy this problem by examining the ways teens talk about their use of new media to craft, pursue, and end intimate relationships. In the first section we trace the practices of contemporary teen courtship and its relationship to the "domestication" of technology, or the way technology defines and is defined by those communities of which it is a part (Hijazi-Omari and Ribak 2008). Teens' stories revealed a set of norms about new media use and intimate relationships. According to most of the teens we talked with, it is appropriate to meet people offline and then pursue the relationship online; if one does meet someone, one should meet that person through friends; one should proceed slowly as he or she corresponds online using the appropriate communication tool; and when breaking up, one should do so in person, or at least over the phone. In the second section we discuss some of the emergent themes about relationships and technology we see from our interviews and observations.

Youth Courtship: Meeting, Flirting, Going Out, and Breaking Up

Liz and Grady, white sixteen-year-olds, sat at the dining room table during our interview, in Liz's family's comfortable middle-class suburban tract home, explaining the role that MySpace played in the origin of their relationship. Grady said that he developed a crush on Liz during the past year, and while he had known her since freshman year, flirting with her in person felt daunting, because, as he put it, "they didn't really talk." Luckily, because they shared a mutual friend, Liz said of her MySpace, "I had him on my Friend list from freshman year ... and that's how you can be friends, just because your friend knows this guy and you kind of hung out with them, so you're like, 'Okay, I'm going to start talking to you.'" Grady used this loose friendship on MySpace to his advantage: "When I had a crush on her, I made sure I talked to her first in class before I sent her a comment on MySpace." Grady carefully planned his first comment to be casual: "My first comment to her was 'Oh, wow, I didn't know we were Friends on MySpace,'" though of course he knew full well they were Friends. After trading flirtatious messages online, they began dating. Liz and Grady are a fairly typical example of the role new media can play in meeting, flirting, and going out. As Grady put it, it is "easier to talk to them [girls] there" than in person, because one can manage vulnerability through what Christo Sims (2007) has termed a "controlled casualness." Indeed, their process is paradigmatic of teens' contemporary meeting, flirting, and dating practices, in which they can pursue casual offline acquaintances as romantic interests online.

Teens have told us that certain technologies and certain mediated and nonmediated practices are more appropriate for certain types of relationships or relationship stages than are others (Sims 2007). As Christo Sims found in his study, "Rural and Urban Youth," in the initial getting-to-know-you part of a romantic relationship, the asynchronous nature of written communication (private messages and comments on social network sites, text messaging, and the more synchronous IM) allows for slower, more controlled intimacy exploration and development. If a given relationship intensifies (because certainly not all flirtatious relationships do), couples typically shift to phone calls, text, IM, and in-person conversations. Social network sites play an increasingly larger role as couples become solidified and become what some call "Facebook official." At this point in a relationship, teens might indicate relationship status through ordering Friends in a particular hierarchy, changing the formal statement of relationship status, giving gifts, and displaying pictures. Youth can also signal the varying intensity of intimate relationships through new media practices such as sharing passwords, adding Friends, posting bulletins, or changing headlines. When relationships end (for those that do), the public nature and digital representations of these relationships require a sort of digital housecleaning that is new to the world of teen romance, but which has historical corollaries in ridding a bedroom or wallet of an ex-intimate's pictures. In the following section we trace the different types of teen courtship practices and the role of new media in these practices.

Meeting and Flirting

As Grady and Liz's story indicates, digital communication often plays a central role in casual relationships and the early stages of serious relationships. New media have provided a variety of venues for teens to meet and/or further potential romantic interests. Instant messaging, text messages, and social network messaging functions all allow teens to proceed in a way that might feel less vulnerable than face-to-face communication. These multiple lines of communication allow teens to follow up on casual meetings or introduce themselves to someone with whom they have only loose ties, perhaps sharing a mutual friend on- or offline. At present, teens' normative practice is not necessarily meeting strangers online (though that does happen) but rather using these mediated technologies to get to know the friend of a friend or further get to know someone with whom one has had only a casual or brief meeting.

For teens interested in someone they may not know well, the plethora of publicly accessible information on a given individual provides a fresh way to "research," or get to know, those on whom they have a crush. Melanie, a white fifteen-year-old from Kansas, in danah boyd's study "Teen Sociality in Networked Publics" said that she does not "talk to people I have a crush on, but I did look up the Honduran twins in our class. We looked at their MySpace." Like Melanie, a teen can research a crush's interests, likes and dislikes, friendship circles, and online behaviors through his or her publicly available social network profiles. John, a white nineteen-year-old college freshman in Chicago, disclosed that instead of asking for a phone number he will "Facebook stalk them" to discover more, though possibly superficial, information about a girl he has met briefly but finds interesting. Much like teens may have historically researched potential love interests through their friendship networks, contemporary teens have additional new media tools for laying the groundwork for flirting and relationships.

After an initial meeting and possible research on their object of affection, teens often use a social network site or an instant-messenger program to intensify a relationship or get to know another person better. This is what adults might think of as flirting or what teens sometimes call "talking" or "talkin' to" (Bogle 2008; Pascoe 2007a). After an initial meeting, a teen might initiate this "talkin' to" by following up through digital communication. As Sam, a white seventeen-year-old from Iowa, said, "The next step, I guess, in this situation is wall posts[3] [on Facebook]—that's kind of less formal. ..." (boyd, Teen Sociality in Networked Publics). Sam noted that if he liked a girl he would post "stupid flirty stuff just trying to make her laugh or whatever through Facebook." At this point, teens flirt, proceeding cautiously, indicating that they like each other, trying to gauge the other's feelings while simultaneously not showing too much earnestness.

The asynchronous nature of these technologies allows teens to carefully compose messages that appear to be casual, a "controlled casualness." John, for instance, likes to flirt over IM because it is "easy to get a message across without having to phrase it perfectly" and "because I can think about things more. You can deliberate and answer however you want." Like John, many teens said they often send texts or leave messages on social network sites so they can think about what they are going to say and play off their flirtatiousness if their object of affection does not seem to reciprocate their feelings. Bob, a white nineteen-year-old[4] living in rural northern California, says he carefully edits his grammar and spelling to give the appearance of an "off-the-cuff" comment. These kinds of deliberately casual messages are evidence of what Naomi Baron (2008) describes as the "whatever theory of language" supported by online communication, in which people are increasingly using more informal linguistic forms to write and communicate. It is important, however, to recognize that these forms of literacy are not a "dumbing down" of language but a contextually specific literacy practice, acutely tuned to the particulars of given social situations and cultural norms.

For example, youth use casual online language to create an intentional ambiguity. From the outside, sometimes these comments appear so casual that they might not be read as flirting, such as the following wall posts by two Filipino teens—Missy and Dustin—who eventually dated quite seriously. After being introduced by mutual friends and communicating through IM, Missy, a northern Californian sixteen-year-old, wrote on Dustin's MySpace wall: "hey.. hm wut to say? iono lol/well i left you a comment ... u sud feel SPECIAL haha =)."[5] Dustin, a northern Californian seventeen-year-old, responded a day later by writing on Missy's wall: "hello there.. umm i dont know what to say but at least i wrote something ... you are so G!!!"[6] Both of these comments can be construed as friendly or flirtatious, thus protecting both of the participants should one of the parties not be romantically drawn to the other. These particular comments took place in public venues on the participants' walls where others could read them, providing another layer of casualness and protection.

Generally, though not always, teens prefer to flirt with people online that they or their friends know or have at least met offline. A minority of teens we interviewed find meeting potential romantic interests online no different from meeting or flirting with attractive strangers they might meet in public, but the general sentiment was that meeting people only online was "weird," "unnatural," "geeky," or "scary." Ellie, a first-year student at the University of California, Berkeley, and respondent in Megan Finn and her colleagues' "Freshquest" study, described her best friend's meeting of her boyfriend on MySpace as weird: "It was really weird at first. She didn't want to tell anyone because she thought it was weird too. But they had such a strong connection that they thought they should meet. And now they're going out." Grady, Liz's sixteen-year-old boyfriend, said something similar about meeting girls online: "I'm not going to start a conversation with a girl on MySpace or text messaging. I'm going to start in person first. Then it's kind of like weird and geeky, you know?" The reasons vary as to why meeting someone online feels weird to some teens. But they all have, in some way, to do with insecurity about authenticity. Brad, a first-year student at University of California, Berkeley, said, "It doesn't seem natural, I guess. 'Cause you're not actually meeting the person face-to-face" (Finn, Freshquest). It is as if that face-to-face meeting allows one to verify who that other person is before embarking on a relationship with him or her.

If teens do meet initially online, they might use their offline friendship networks to verify the authenticity, safety, and identity of the person with whom they are corresponding. Dana, a Latina fourteen-year-old from Brooklyn, New York, met her boyfriend online through mutual friends. Her best friend's boyfriend's best friend saw her MySpace and

> he requested me 'cause he liked what he seen, and then my best friend talked to him about me and then because of MySpace we were goin' out. ...' Cause if MySpace wasn't there, then I woulda not had him as boyfriend. We talked on AIM and then we exchange the numbers, and then I met him. I seen him before, but I got him noticed on MySpace and now we're together. (Sims, Rural and Urban Youth)

Like other teens we talked to, Dana and her boyfriend flirted online before they moved to offline communication and eventually met in person. Dana said, "He usually started getting on AIM every day, and I started talking to him from there." They communicated for two days through MySpace until they traded phone numbers and "talked like from twelve to six in the morning." Eventually they met in person in a public space—a local park—in the company of groups of friends. Dana's story is not an uncommon one. Teens regularly meet romantic interests through shared friends in online environments, using these online networks to further offline meetings or deepen casual ties to online friends. Teens rely on their networks to do some of the verification work in these online settings.

Though the in-person meeting went well for Dana, other respondents expressed hesitancy about moving online relationships offline for fear that people might not live up to their online personas. John, the Chicago freshman, asked, "What happens after you've had a great online flirtatious chat ... and then the conversation sucks in person?" He experienced this phenomenon firsthand as he transitioned from high school to college. John had used Facebook to add as Friends "the girls you wanted to meet before school started that you thought were hot and wanted to get a head start on." However, once he reached his university in the fall, "you actually saw them and didn't say anything … the game was over." When asked why he didn't talk to them in person, he said, "You didn't say anything, because what are you gonna say … 'Hey, you're my Facebook Friend?' The key is to meet them in person … then Facebook them." Brad, the Berkeley freshman, expressed similar hesitancies about meeting people offline. "You don't know that's who you're meeting. It isn't a smart thing. And you'll end up idolizing the person, thinking they're just this perfect thing. But they probably aren't because no one is perfect. And it's just a big letdown." This "hyperpersonal effect" indicates that intimacy might be heightened online in a way that might not translate seamlessly into offline relationships (Walther 1996).

While most teens express hesitation about meeting people online, in the case of marginalized teens, the Internet allows them to meet other people like themselves (Holloway and Valentine 2003). This sort of digital contact provides a means for youth who didn't feel heard or who felt otherwise disenfranchised in their communities to participate in other ways (Maczewski 2002; Osgerby 2004). For example, Gabbie, a seventeen-year-old first-generation ethnically Chinese teen from California, wanted to find a Chinese boyfriend, but potential suitors were in limited supply in her immediate community. In part because of this desire, she joined the social network site Asiantown.net and struck up communication with a young man she found attractive. "Well, right now I'm talking to this guy. But he has a girlfriend. I don't know. We're just talking as like friends. It seems like he's being a little flirty, but then … I don't know." The boy she is talking to lives in the Central Valley, about an hour from where Gabbie lives. We rarely heard teens such as Gabbie, who lack specific offline social circles, talk about moving these relationships offline as being unnatural or weird.

In a similar way, new media also are important tools for gay teens who want to date, because "the biggest obstacle to same-sex dating among sexual minority youth is the identification of potential partners" (Diamond, Savin-Williams, and Dube 1999, 187). It allows them to meet other teens for friendship or dating and affords them a level of independence, as it does for straight teens, to carry on relationships outside the purview of their parents if need be (Hillier and Harrison 2007). The Internet can put gay teens in touch with other teens so that they can have the romantic experiences that their heterosexual counterparts presumably find more readily in offline contexts. Robert, a white seventeen-year-old at a private school in Chicago, became so frustrated about not finding other guys to date through his offline friendship circles that he wrote a Facebook "note" about his difficulties dating as a gay teen:

> Every time I have a crush or something, it doesn't work out (he's not gay, not enough time, etc). I'm not a downer, but I'm just realizing that if a straight person's chance of compatibility is 1 in 100. AND only about 3 in 100 are gay, and the compatibility is still 2%, then my prospect is .03 in 100, or 3 in 10,000. That is not very encouraging!

Robert said that a friend set him up on a blind date as a direct result of the announcement he placed on Facebook: "Andrew, another gay guy at my school, and [my] friend, set me up with Matt because he saw my desperate note on Facebook!" Matt and Robert were introduced through Facebook and after the initial setup, Robert was giddy with excitement and said, "We've been texting the past few days a lot; he is really good looking, and a jock, believe it or not, but we seem to really have hit it off. I hope for the best." The two had a very sweet day picked for their first date: Valentine's Day. Much like Dana, Robert found a date through a shared friend. But unlike straight, more mainstream teens, he expressed no hesitancy about meeting in person someone he had met online.

Going Out

Technology also mediates teens' long-term, steady, and committed relationships. Teens in relationships have high expectations of contact with and availability of their significant others as well as expectations that the relationship will be publicly acknowledged through digital media. These expectations of availability are compounded by the "always on" (Baron 2008) possibilities of new media. Additionally, these media help teens reach out beyond their institutional constraints, allowing them to maintain romantic relationships their parents wouldn't necessarily approve of as well as sustain relationships that might be geographically challenging. Like Jesse and Alice, introduced at the beginning of this chapter, teens who are steadily dating frequently text or call each other, post pictures of each other on their social networking sites, rank order their Friends in a particular way, and exchange digitized tokens of affection, signaling to their significant other and their online publics that they are in a relationship.

Being in a relationship increases expectations of availability and reciprocity, which has implications for how teens use new media, given this "always on" potential. In practice this means that youth in a relationship exchange several phone calls, texts, and/or IMs a day. Teens use this intensified contact as a way to differentiate romantic relationships from other relationships—to indicate that their relationship is special or different. Zelda, a Trinidadian American fourteen-year-old from Brooklyn, New York, explained that if one is in a relationship and doesn't respond to a message, the other person will "probably get mad. If they call you and you don't pick up, they probably get mad. If they write a comment on your page you have to comment them back" (Sims, Rural and Urban Youth). He distinguishes this from interacting with friends through digital media: "It's not like it's a normal friend; it's your girlfriend or whatever. You're in a relationship; you're not supposed to just answer whenever you want." As noted in chapter 2, youth have expectations of reciprocity in online communications, and these are heightened in intimate relationships. Teens now do much of their relationship work by using new media—reciprocating in comments, differentiating their romantic attachments from less intimate friends, and giving priority to phone calls from significant others.

To signal to each other that they care and are in an intimate relationship, teens exchange small digitized symbols of affection, much like teens in the 1950s traded rings, jackets, or bracelets. Champ, a nineteen-year-old Latino who also lives in Brooklyn, explained, "Like if she's already your girlfriend, you probably send a little text message, 'Oh I'm thinking of you,' or something like that while she's working. ... Three times out of the day, you probably send little comments" (Sims, Rural and Urban Youth). These comments are the digital interactional work that cements contemporary teen relationships. Derrick said,

> You know in your head you've just got to do it. It's like she writes you a comment; write her a comment back. It's not like a friend thing. It's not like your homeboy just wrote you a comment like "oh man, this kid wrote me a comment again." Write her a comment back. (Sims, Rural and Urban Youth)

Youth do emotional work to maintain a relationship through digitized media. Rather than (though sometimes in addition to) love notes exchanged in between classes, youth demonstrate affection through private and public media channels.

These tokens are part of the interactional relationship work that happens through new media; another is the expectation of availability. Teens find that their significant others expect frequent check-ins, usually by mobile phone. Derrick said,

> When you're in a relationship one thing I learned [is] always pick up the phone for your girl because she complains if you don't. ... The thing about a cell phone when you're a teenager is if you have a cell phone and you don't pick it up you're doing something that you're not supposed to be doing. (Sims, Rural and Urban Youth)

As Christo Sims notes in his research on urban and rural teens, teenagers are expected to account for their whereabouts. They are beholden to parents in this sense but also to significant others, especially in relationships in which trust might be missing or weak. As a result it might be hard to preserve space or time for oneself outside this frequent contact. In fact, Zelda said he knows he needs to answer the phone regularly because if he doesn't, "they probably going to get mad" (Sims, Rural and Urban Youth). The phone especially acts as a sort of leash, a way to keep tabs on a significant other, much like parents keep track of their teens. Teens seemingly endure this leash because of the increased independence afforded them by the phone.

In addition to the expectations of regular, if not continual, contact, teens affirm and are expected to affirm their relationships online, both by and for their significant others and for their networked publics. Zelda underscored the importance of representing relationships online: "You gotta acknowledge on your page that you [are] like with her" (Sims, Rural and Urban Youth). They define and affirm their relationship status, give public tokens of affection, and post pictures. On Facebook, default relationship options are preset, so in addition to indicating an "official" status, teens have creatively developed ways to include nuance and detail in their relationship descriptions. The existing categories hide a variety of relationships and elide the depth or length of a given relationship, so teens sometimes remedy this by indicating the seriousness of a particular relationship through noting its duration, a particularly popular practice among youth interviewed by Christo Sims in Brooklyn, New York. According to Dana, mentioned earlier, couples write a relationship-origin date in their MySpace headline "to show that they have a relationship or something, so like that's showing more, and it shows that he's in a relationship" (Sims, Rural and Urban Youth). The statement of a relationship anniversary is both a signal of intimacy to one's significant other and a hands-off signal to other teens who might be interested in one member of a couple. Nini, a Latina thirteen-year-old from Brooklyn, said,

> If you put the relationship date, whenever you got together, the girls know that you're in a relationship and this is the date, so don't really get into it with the boyfriend, 'cause you are really falling for each other … they know that you're a year, so I'm not gonna mess with the boyfriend. (Sims, Rural and Urban Youth)

Nini highlights the "hands off" message, arguing that the length of time a couple has been together indicates the intensity of their relationship to potentially meddlesome outsiders.

Couples typically negotiate offline the act of putting their relationship status online, whether it be a simple "in a relationship" status on Facebook or a more nuanced relationship date on MySpace, notes Christo Sims. Teens dismiss the practice of posting these sorts of public notifications about changes in their relationships through online venues before discussing it with their partner first, usually offline. Joan, a first-year student at the University of California, Berkeley, said,

> Yeah, I have friends [who] have confirmed they have gone official with their boyfriends through Facebook, which is ridiculous. I have known people that are dating and they'll get a request "so and so said that you are their girlfriend." They pushed the button and they are like, "Oh my God, we're official." (Finn, Freshquest)

Teens seem to have the sense that this sort of intimate decision should be made interpersonally, not just announced digitally.

The whole of these social network profiles, not just the relationship status, are the digital embodiment of teens' relationships. When in a relationship, teens rank their Friends to indicate the seriousness of their commitment. Derrick said that "you probably write something, have her on your Top Friends, don't put other girls, don't have girls write messages to you saying anything crazy. Just to make her feel better" (Sims, Rural and Urban Youth). When teens in a relationship do not rank their Friends in a way that reflects their relationship status—that is, they do not rank their significant other high among their Friends—conflict might result, as it did with Jesse and Alice. Jesse confessed, as he showed off his MySpace site, "Alice was actually not my original top one." Alice paused from her needlework to jump in the conversation and said, indignantly, "I was like number twelve or something." Jesse, clearly defensive, his voice growing higher, cried, "Does it really matter? You know! Really? My number one? Really?" Alice responded a little sarcastically, rolling her eyes, "Like he's not number one on my account." Clearly, it was not the first time they had had this discussion. While for these two teens, the tension did not challenge the basic foundation of their relationship, their disagreement indicates how important these public representations of relationship intensity are. Alice's feelings were hurt by Jesse's refusal to place her above his other Friends on his list.

In addition to ranking Friends, youth in relationships need to leave public messages for and post pictures of their significant others. Doing so sends messages to their significant others about their dedication and to their digital public about the nature of the relationship. Zelda said, "Sometimes, like on MySpace, you will leave a comment, and you leave a whole bunch of stuff on there 'cause they your girlfriend and stuff, so everybody can see your name. Girls get happy for that. I don't know why. They just get happy" (Sims, Rural and Urban Youth). Zelda explained that he comments "on their pictures. Like if they got a new picture up, leave a comment 'oh, that's a nice picture you got up' or whatever." Zelda indicated the dual message contained in this sort of digital relationship work—the girlfriend is happy because this sort of work feels attentive and loving and Zelda sends a message to their community, "everybody can see your name," about his dedication to his relationship. Another form of relationship work includes posting "couple" pictures on one's social network profile. As Derrick says, "Throw a picture in there of her on your profile. Have it in your pictures like when people look at your pictures they see you and her together or something. Something that makes her say, 'Aaahhhh.' To show her that you care for her" (Sims, Rural and Urban Youth). Again, Derrick's comment shows that these tokens are both for a significant other and a teen's audience. These practices also hold members of a couple publicly accountable. Once one states that she or he is in a relationship, this insures that both members of a relationship agree on their status and are ready to make it public, thus prohibiting one member of the couple from arguing that "it wasn't official."

Breaking Up

Because of the integration of new media into their relationships, teens also experience mediated breakups. These new communication practices often require that teens take a variety of steps to sweep up the digital remnants of a given relationship and to deal with access to and the continuing digital presence of their former significant others.

The media that some youth laud as a comfortable way to meet and get to know a romantic interest are viewed as a poor way to break up with an intimate. Billy, a white seventeen-year-old from a northern California suburb, said that as he was IMing with a friend he advised his friend to break up with his girlfriend. Apparently his friend did so right then, through IM. Representative of other teens, Billy said, shaking his head, "That was bad." Grady, Liz's sixteen-year-old boyfriend, agreed that breaking up through IMs or text messages was "lame" but that only "some people do it; most people don't." In line with a theme we heard, Grady claimed that breaking up in writing either through a social network site or through a text message was "disrespectful. Because they can't say anything back or anything." Teens acknowledge that breaking up in person is preferable to using text messages, instant messages, or messaging functions on social network sites, because face-to-face interaction is more respectful. Just as teens are thankful for the ways in which they can manage vulnerability using new media in the early stages of relationships, they sense that this vulnerability should not be managed in the same way at the end of a relationship.

New media have created a public venue for digital remnants, where digital representation might outlast the relationship. For instance, Gary, a seventeen-year-old Filipino senior from northern California, had created his MySpace site with his now ex-girlfriend. He laughed sheepishly during an interview as he logged on to his profile and the site title bore both his name and that of his ex-girlfriend, reading, "Sarah will always love Gary." This passive digital residue of their history together remained long after the relationship was over. Even though teens say that the actual act of breaking up should not happen in a mediated way, breakups do take place online as youth sweep up the digital remainders of their relationships. Teens' breakups can be reflected passively through status changes or displayed actively through hostile public messages and announcements. Michael and Amy exemplify an actively public breakup—public animosity, angry messages directed specifically at an ex-intimate, and the seeking of public validation from their friends. Conversely, passively public breakups entail quietly removing pictures, changing one's relationship status, and reordering Friends. While these breakups also happen in public, they are tamer and perhaps more representative of the customary way teens end relationships. Trevor's most recent breakup exemplified this passively public practice. The white seventeen-year-old from suburban northern California said that he usually places the person he is dating as the top Friend on his MySpace and moves people instantaneously when they break up. But "the latest ex stayed on there for six months because I was waiting. … I thought I'd be in a relationship really quickly." Trevor says that his ex-girlfriends weren't upset when he removed them. "There was never drama about it. They got it. They understood. … I always try for that, because I really don't want to be the jerk." For teens, changing a public representation of a relationship is a normal part of these now-mediated relationships; thus, unless the couple does not agree on the status of their relationship, they are rarely surprised by this sort of alteration of an ex's profile.

After a relationship ends, teens often inhabit the same, or overlapping, networked publics. Frequently, members of a former couple can still see each others' profiles, see messages left by their ex–significant other on shared Friends' social network profiles, and receive automatic updates about their ex, should they retain him or her as a Friend. As Christo Sims's research has highlighted, these indirect communication channels mean that youth can still be in touch with and possibly monitor each other after an intimate relationship has ended. These communications can be caring, respectful, retaliatory, hurtful, or angry, or they can be ways to send messages to an ex–significant other without having to interact directly with him or her. While teens may have the sense that they should sever real-world and digital ties with their former girlfriends or boyfriends, Bob, the white nineteen-year-old from suburban northern California, said that monitoring one's ex on a social network site is

> one thing that you shouldn't do but everyone does. You can go check all their stuff. Like you look at their Facebook, you look at their MySpace, you see if they take off the photos of you, you see if they changed their relationship status to something, you see if they've got a new person writing on their wall. Like you become a stalker, and a highly efficient stalker. Because all the information is already there at once. You don't have to ask your friends or her friends if she's seeing someone new. Like you know. And then they want you to know. (Sims, Rural and Urban Youth)

Teens are aware that their exes see them in these networked publics and use the opportunity to communicate with them, though not directly. Ono, a Haitian American sixteen-year-old from Brooklyn, New York, used the opportunity provided by social network sites to communicate her anger toward her ex-boyfriend.

> You want to make them feel so bad that the relationship ended. So you take out all the comments, unless, it depends, unless you are still friends with that person. Take out all the pictures. Put some other person, or maybe delete him from your Friends list, and, but you know that he's gonna look at your profile anyway, so you put other males next to you, or put pictures of another male and say how nice he looks in that outfit or whatever, or my future man, or whatever, so you could put as much anger in that person as you can, or if you guys have the same Friend, like if me and my boyfriend have you as a Friend, I'll use you to get his attention. (Sims, Rural and Urban Youth)

Ono strategized about how to use her shared public to make her ex-boyfriend feel bad by signaling that she had severed ties with him, that he was no longer her Friend, and that she was intimately connected to other boys. The same technology used to publicly affirm intimate relationships can be used to publicly demonstrate their demise and to communicate anger toward someone with whom a teen may no longer have direct contact.

Bob used the same technology to communicate to an ex-girlfriend a gentler message. He had just endured a "really rough breakup" with a girl who wanted to "get back together" with him, though he did not reciprocate her wish to reunite. He wanted to communicate to her the fact that he was not willing to reconcile, but he felt constrained because he had learned of her desire in confidence from a mutual friend. To communicate his feelings to her, he changed his relationship status on Facebook to "in a relationship," even though he was not involved with anyone. At that point his ex-girlfriend realized that "I was unavailable. I knew she would read that; I didn't tell her or anything, but I knew that she would find it. And so that ended it officially." His ex-girlfriend communicated back to him in a similarly passive way:

> I go on her MySpace and there's a blog about how she can finally move on. But it's addressed to no one. Right? I know who it's talking about; she knows who it's talking about. So that was a weird instance where "I'm not telling you but I know you're going to find this." (Sims, Rural and Urban Youth)

These sorts of indirect communications can enable teens to exit relationships in a dignified way and enable them to "have their say." Instead of communicating through oral conversations, or less directly through handwritten notes or chains of friends, teens can passively communicate through their online profiles and presence.

Despite popular emphasis on the one-to-one communication opportunities provided by these technologies, youth often use them to communicate indirectly, both through the technology and through intermediaries. Christo Sims's research on the ends of relationships shows that through new media, teens can retain an indirect channel to communicate after breaking up. While teens stop engaging in continuous contact after a breakup, they still use new media to communicate indirectly with each other and their larger mediated publics. Mechanisms on social network sites for indicating status or posting to an undefined public enable teens to delegate some of the more awkward social articulation work to technology-based, mediated forms of communication.

Intimate Media: Privacy, Monitoring, and Vulnerability

Themes of privacy and vulnerability weave through teens' new media practices. The ability to monitor one another and be monitored, emotional and physical vulnerability, and tensions around privacy thread through the variety of intimacy practices in which teens engage. Digital communications allow teens a sphere of privacy, when they don't have their own spaces, to communicate with their significant others through a circumvention of geographic and institutional constraints. The ability to talk beyond the earshot of one's parents and other adults, such as teachers, is part of this circumvention. Teens told us that the ability to communicate outside of adults' view and hearing was important. For instance, Joan, the Berkeley freshman, claims that she and her first boyfriend would talk

> online all the time, all the time. Like, we talked on the phone but then sometimes we talked on the phone and IMed at the same time … especially it's like our parents was in the room and then we would talk to them and then if there is something that you don't want your mom to hear you could type it and then you could talk about it. (Finn, Freshquest)

Similarly, youth are able to maintain relationships with people of whom their parents might not approve, much like Jesse and Alice, because of this privacy. However, given the expectations of high contact with other teens and the amount of personal information in a semipublic realm, teens also have to negotiate new boundaries and spheres of privacy in their intimate relationships (Livingstone 2008). In this sense, social media carve out a new private realm in which teens can communicate, largely outside the purview of adults, while simultaneously redrawing and often weakening boundaries around their personal spheres of privacy.

Monitoring and Boundaries

From investigating crushes, to being in contact with significant others, to enduring breakups, the aspects of digital media that let teens be constantly in touch also allow them to monitor one another more intently. This monitoring varies from researching potential love interests to using a shared password to check up on one's significant other to attempting to restrict one's significant other's communications with his or her friends. Some youth regularly check on their significant other's websites simply to see what they are up to. Gabriella, a Latina fifteen-year-old from Los Angeles, logged on to her boyfriend's profile daily as part of her routine after she logged on to her own, "just to check" (boyd, Teen Sociality in Networked Publics). Similarly, Samantha, a white eighteen-year-old from Seattle, admitted, "I have done some checking up [on my boyfriend]" (boyd, Teen Sociality in Networked Publics). This sort of "checking" behavior happens when one has a crush, when one is monitoring one's romantic partner, and sometimes after a breakup.

The importance of passwords to one's online presence is central to these monitoring practices. Sharing a password both denotes intimacy and allows a significant other to monitor the private portions and manipulate the public parts of a social network profile. For some couples, such as Clarissa and her girlfriend, Genevre, white seventeen-year-olds in northern California, sharing a password feels like a way to maintain a connection even when they are apart. In fact, as Clarissa logged on to her MySpace profile she laughed, seeing that her girlfriend had updated it and altered the background to a more attractive one. However, not all teens feel comfortable with the amount of power a significant other wields with the password. Derrick, the Dominican American sixteen-year-old living in Brooklyn, New York, argued that girls want the passwords because

> they want to check up on you all the time. They want to get your MySpace password, they want to get your AIM password, they want to get your phone, your answering machine, the password. They want to get anything they … know that another girl can get in contact with you through. (Sims, Rural and Urban Youth)

While Champ, the Latino nineteen-year-old from Brooklyn, shares his password, he protects his privacy by changing his password regularly. "You gotta change it. … I'll be changing mine like every three weeks" (Sims, Rural and Urban Youth). Clarissa's, Derrick's, and Champ's varying responses to sharing a password show how this practice is both a sign of intimacy and a possible invasion of privacy. By refusing to share it, some youth attempt to set a boundary around their intimate relations, sometimes to the frustration of their significant others, usually girlfriends. This may be because some girls feel powerful when they know their boyfriend's password. Dana, the Latina fourteen-year-old living in Brooklyn, explained, "I made my boyfriend give me his password and that shows power" (Sims, Rural and Urban Youth). Given the research that documents continuing gender inequality in heterosexual adolescent dating relationships (Hillier, Harrison, and Bowditch 1999; Hird and Jackson 2001; Jackson 1998), it is not surprising that girls are strategizing ways to feel more powerful in these partnerships.

In a similar move, some of the youth we spoke with draw boundaries by altering digital footprints that might make their significant other question their commitment. These footprints may be messages, search histories, phone numbers, or texts that reveal one's intimacy practices to families, siblings, friends, or significant others. Zelda, the Trinidadian American fourteen-year-old living in Brooklyn, New York, actually deletes information on his site to get rid of evidence that might anger his girlfriend: "Sometimes I'll just go in there and I delete stuff that girls wrote me. I'll just delete it." To avoid these privacy compromises, Champ and Zelda change the names on their mobile phones. To prevent his girlfriend from scrolling through to look at his contacts and call logs, Champ records "their names different," explaining, "Yeah, if it's a girl's name, you put a boy's name that probably sounds similar to it. … Like, let's say the girl's name is Justine, you'll probably put Justin" (Sims, Rural and Urban Youth). While these technologies have provided a greater realm of privacy, digital footprints might compromise this privacy and thus youth are often drawing digital boundaries to protect a personal sphere.

Some of the monitoring that happens during teens' relationships veers eerily close to serious emotional control or abuse. Lolo, a fifteen-year-old Latina from Los Angeles, said that her boyfriend did not like the fact that her social network profile was public. Using the password she shared with him, "He kinda put it on private, hello. He's like, 'I don't wanna know every boy's going in there searching you'" (boyd, Teen Sociality in Networked Publics). We heard this insecurity over their claim on their romantic partners throughout our interviews with youth. Teens may intensify some of the monitoring practices we found as a way to attempt to control some of their anxiety about the stability of their relationships.

This sort of control might also intensify when economic transactions are involved. In our research, teens sometimes paid their own phone bills, but usually their parents paid. This meant that teens needed to obey their parents' rules (to the extent the parents could enforce them) about mobile phone use. Something similar happened when one's significant other paid the phone bill. Ono, the Haitian American sixteen-year-old living in Brooklyn, New York, said that her friend's boyfriend pays her friend's phone bill and as a result

> he expects when he calls, even if she's not available, to just pick up and say, "I can't talk to you right now, I'll call you back." Or if he's with her, then he would be asking who else is calling if it's not her parents or something. That's what happens when he pays your bills. And yeah, he can talk to you every day, even if you're not free, because he pays for it. (Sims, Rural and Urban Youth)

Girls in this type of relationship seemingly trade one type of control, parental, for another (Hijazi-Omari and Ribak 2008). Their privacy is compromised because they do not retain economic control of their mobile phones.

Youth monitor one another in the early stages of, during, and after the ending of the relationships. This monitoring manages anxiety so central to teen relationships in which teens for the first time are crafting intimate ties with one another. The monitoring capabilities afforded by digital media seem like a way to manage such anxiety as teens seek to put to rest their fears about vulnerability and betrayal. The ability to monitor others through these new media venues both allows teens to learn about others and makes them vulnerable to surveillance and control by others.

Vulnerability

New media simultaneously increase teens' vulnerability and their control over their emotional exposure. This heightened vulnerability may allow teens to craft new and strong emotional connections with one another as well as render them more open to being victimized by their friends, acquaintances, and other adults. However, the removed and asynchronous nature of some new media also allows them to manage emotional exposure and render teens less vulnerable, especially in the early stages of a relationship.

Boys in particular, because of contemporary association of vulnerability with a lack of masculinity (Korobov and Thorne 2006), express relief about the extent to which new media allow them to control what they perceive as emotional vulnerability. They feel less exposed because they can text a girl or leave a message on her MySpace page rather than risk embarrassment by calling her and stumbling over their words or saying something embarrassing. Bob, for instance, said,

> It's a lot easier to flirt digitally than it is in person 'cause there's no awkward silence. You can't say something you don't mean 'cause you could sit there at one comment on a person's profile and spend a half an hour making sure that everything is right. Like some words are lowercase on purpose. The punctuation's just the way ... I want it to look sloppy, but it really has this, you know, acute meaning to it. (Sims, Rural and Urban Youth)

The asynchronous nature of texting and leaving messages allows boys to save face when flirting with a new girl. In this way, the controlled casualness discussed earlier is a form of emotion management and a way to control vulnerability.

The same technologies that allow youth to manage emotional exposure might also render them more vulnerable, in part because of the amount and type of information shared and the speed at which it can travel. Teens are not necessarily in control of digital representations of intimate practices or in control of the audience who sees those representations. For instance, Elena and Brett, two gregarious white sixteen- and seventeen-year-olds, respectively, from northern California, talk about how embarrassing pictures might be forwarded. Elena said, "That's a lot of drama too. They can send pics to other people." Brett continued, laughing, "People might take a picture of other people making out at a party." Elena continued, "Like so-and-so was kissing so-and-so or that so-and-so made out with so-and-so at a party. Then the next week they're like, 'Look at the picture; obviously it meant something.' Then they're with somebody else." Elena said that the picture might get "around school and you're like, 'Wait, how did you even get this picture? You weren't even at the party.' It goes further than you think sometimes." In this way, even teens' offline practices may be monitored online if people forward compromising pictures of them. This digital proof of one's intimate life may spread rapidly, outside of one's control.

The other vulnerability teens talked about is that of the stereotypical risk conveyed through fear-based narratives of the Internet, that of the stalker, the stranger, and the predator. Teens rarely mentioned these stories in our research (apart from noting that this was what adults were concerned about), but a minority of youth reported having negative interactions with predatory-type adults online. Those youth who seek out intimate communities online, such as gay teens, might be more at risk for this sort of unwanted stranger intimacy. For all the opportunities to create community for gay teens, the Internet also puts them at risk as they seek this community. Robert, the white seventeen-year-old from Chicago, told a particularly affecting story about his experience on the Internet as he was coming into his early teens.

> A couple times a week, after my parents went to bed, I visited some Internet sites … then after a while, I found a chat room website, a gay teen chat room. I chatted with a lot of guys; eventually I started to talk to people outside of the chat room, on MSN Messenger. There were people who wanted to do things with cameras and pictures, and for a while I went along with some of it, not really doing too much. Then one day, it wasn't a teenager who sent me their pic, but an old fat man. I was disgusted, beyond words. I smashed my computer camera, deleted my MSN, and barred any memory from those times out of existence until I recollect now.

Robert was trying to explore his sexuality the best he could, as a single gay teen, but in doing so, he ended up on non-age-graded sites, which, though not inherently risky or problematic, may be dangerous for marginalized teens looking for community. Instead of getting to experiment in more public and socially acceptable ways, through structured rituals of heterosexuality, gay teens often must find their own way. On the one hand, the Internet is an invaluable lifeline, but on the other, it renders gay teens more vulnerable to situations such as this one.

New media allow teens to manage their vulnerability; permit them to have intensely emotional, vulnerable conversations; and render them potentially susceptible to the forwarding of information about them and vulnerable to those who wish to take advantage of them.

Conclusion: Controlled Casualness, Continuous Contact, and Passive Communication

While many adults may perceive social network sites as being simply glorified dating sites, this chapter, in conjunction with chapter 2, on friendship, demonstrates that teens are not one-dimensional beings interested only in prurient communications and subjects; rather they craft complex emotional and social worlds both publicly and privately on and offline. Academic work has rarely taken youth courtship practices seriously, but in examining the way teens talk about these practices and their emotions about them, our project demonstrates that romance practices are central to teens' social worlds, culture, and use of new media. For contemporary American teens, new media provide a new venue for their intimacy practices, and render these practices simultaneously more public and more private. Teens can meet people, flirt, date, and break up beyond the earshot and eyesight of their parents and other adults while also doing these things in front of all their online friends. As chapter 2 also points out, participating in these mediated relational and emotional practices is central to being a part of an offline social world. Youth are developing new kinds of social norms and literacies through these practices as well as learning to participate in technology-mediated publics. These sites of peer-based learning need to be taken seriously, as they are structuring social and communicative practices that differ in some important respects from the experiences of these teens' parents, and they can become a site of intergenerational tension and misunderstanding.

When meeting and flirting, teens find online communication extremely useful. This is especially true in terms of furthering casual acquaintances. They have more freedom to get to know friends of friends or others they have met briefly at parties or other group gatherings without risking too much embarrassment. They can also use social network sites to learn about, usually unbeknownst to the other person, someone in whom they have an initial interest, be it someone they see every day in class or the person who sells them burgers at the local fast-food restaurant. While meeting people solely online is not the norm, some teens do meet and flirt that way. Others consider this brave, scary, or weird, depending on their perspective. Their messages and interactions during this time might be characterized as a "controlled casualness." Dating teens use new media often, engaging in what one might think of as "continuous contact." When in a relationship, teens frequently communicate with each other and expect their significant others to publicly acknowledge and maintain their relationship on their social network profiles. Teens' relationships also end in the presence of their networked publics. The breakups might be active or passive, but because of their shared publics, teens retain the ability to passively communicate with each other even after ending intimate ties. Their continuing indirect communication about relationship status is a way in which these sites enable intimate content to be made very public. This publicity both allows teens to exact revenge and communicate important, but indirect, messages about their emotional states to their former significant others. Because of the dearth of research on teens' intimacy practices, we lack comprehensive comparative case studies, but it seems that teens' current use of new media might be a unique moment in the recent history of teen dating practices. New media allow, and seem to encourage, teens to make relationships and relationship talk explicit. They let teens access romantic others' personal information and share versions of or information about themselves that might not be done as easily in offline circumstances. Much as friends have in the past, technology now acts as a social intermediary, enabling communication that is passive, but very important, at liminal relationship stages, such as beginnings or endings. Finally, among teens in relationships, technology allows them to maintain a passive copresence with each other and provides new ways to subvert expectations of that copresence.

As we saw in the case of friendship practices, these online tools and communication practices make peer-based interaction and pressures more consistently available to teens. Unlike more familiar forms of public space, networked publics and private communication channels such as IM and mobile phones can make it harder for parents to passively monitor their children's romantic communications (though written records of these communications often linger in digital environments should parents know how to access them). Youth call and send messages to each other directly, bypassing mediation by parents or siblings. This is part of the trend toward what Misa Matsuda (2005) has called "selective sociality," in which youth can make more intentional decisions about those with whom they affiliate. Further, some parents do not fully understand the norms that govern teens' online interactions, and the literacies they deploy in these interactions, and thus they may be tempted to resort to blanket prohibitions rather than more nuanced forms of guidance. These dynamics are explored further in chapter 4.

The snapshot of contemporary teens' intimacy practices presented in this chapter indicates that today's teens are part of a significant shift in how intimate communication and relationships are structured, expressed, and publicized. Networked publics of different sizes and scales contextualize these intimate communications and practices, allowing youth to observe the intimate interactions of others, and conversely, to display their own emotions, practices, and relationships to select publics. The new possibilities of self-expression available online, characterized by more casual and personal forms of public communication, complicate our existing norms about the boundaries between the public and the private.

Notes

1. As with other parts of teen culture, contemporary practices of dating and romance are deeply gendered (Best 2000; Martin 1996; Pascoe 2007a). Gender difference and inequality is central to heterosexuality and thus is embedded in dating practices.

2. Indeed, in spite of the current flurry of concern over what kids are doing online, the Internet and social network sites have hardly led to an explosion in teen sexual behavior. In fact, the number of teens who say they have had sex before they graduated high school has declined from 54.1 percent in 1991 to 47.8 percent in 2007 (CDC 2007).

3. A "wall" is the place on a typical social network site where someone's Friend might leave a message for him or her to read. These messages are usually visible to others, but their public nature depends on the privacy settings of a given profile.

4. Christo Sims interviewed Bob several times, such that during the course of our research Bob's age ranged from nineteen to twenty-one.

5. Like many teens, Missy wrote using typical social media shorthand. Translated, her comment would read: "Hey, hmm, what to say? I don't know. Laughing out loud. Well, I left you a comment. … You should feel special haha (smiley face).

6. "G" is slang for "gangsta," in this case an affectionate term for a friend.

Shopping for Love

Online Dating and the Making of a Cyber Culture of Romance

By Sophia DeMasi

Ten years ago, a reader of any mainstream national publication or local weekly could not have helped but notice the ubiquitous personal advertisements that saturated their back pages. Today, these same personal advertisements have migrated to the virtual pages of the World Wide Web. A casual glance at the content of online personal advertisements suggests that their writers solicit readers for a variety of reasons, including friendship, a long-term relationship, and casual sex; however, dating dominates the virtual landscape. The vast majority of online personal ads are written by people who want to date as a prelude to a satisfying long-term relationship (Brym and Lenton 2001). With an estimated 2,500 dating websites in operation entertaining approximately 40 million visitors each month (Sullivan 2002), Internet dating is now a popular and vital part of the process by which people seek and find intimate partners.

Until recently, the use of personal advertisements to locate intimate partners was understood as a deviant activity resorted to only by "losers" left out of the marriage market, or "perverts" seeking illicit sexual encounters. In addition, the stigma attached to users of personal ads made many reluctant to reveal the activity to others (Darden and Koski 1988; Rajecki and Rasmussen 1992). Those days are gone. Now it is quite common for people to publicly reveal their experiences with online dating. It is not at all unusual to hear coworkers, friends, and acquaintances shamelessly boast about a successful date or to confess to a dating fiasco that began with an online personal advertisement. In offices, classrooms, and living rooms across the country, online daters boldly relate the latest update to their online profile and how many "hits" it got.

Another sign of the astonishing popularity of online dating is its visibility in popular culture. Currently available are over thirty self-help books that instruct users of dating websites in the finer points of creating an effective online profile. Book titles like *E*

Dating Secrets: How to Surf for your Perfect Love Match on the Internet, Everything you Need to Know About Romance and the Internet, 50+ and Looking for Love Online, and *Worldwide Search: The Savvy Christian's Guide to Online Dating* indicate that online dating now has mass appeal across sexual orientation, race, age, and religious groups. The popular press has also put Internet dating in the spotlight by publishing the revelations of ordinary people who found their companions through online personals (Foston 2003; Wilkinson 2005). Self-help books and the public testimony of online daters help put to rest the belief that online dating is something resorted to only by desperate people.

The tremendous expansion of online personals, along with the public pronouncements of the people who use them, suggests that technologically mediated dating is now a socially acceptable method for finding intimate partners. Stigma and shame are no longer associated with people who seek to connect with others through personal advertisements. How has the formerly "deviant" activity of using print personal ads to seek and find partners given way to the apparently routine practice of seeking and finding companions through online personal advertisements? Moreover, what consequences might this change in medium have on the process of finding romantic and/or sexual partners?

Recent technological innovations and demographic changes are part of the reason why online dating has become such a common practice in the first decade of the twenty-first century. More significant, however, are the deliberate marketing strategies used to increase the appeal of online dating. Together, these factors work to expand possibilities for finding partners and establishing intimate relationships. But, paradoxically, online dating also limits the possibilities for creating relationships, particularly those that exist outside the narrow confines of relationship ideals historically identified with heterosexual intimacy. To attract a large number of participants, dating websites rely primarily on a particular construction of intimate relationships that emphasizes love, romance, and monogamy; they rarely mention sex for pleasure and the desire for physical intimacy. Moreover, embedded in the structure of online dating websites are existing gender and sexual identity categories that preclude explorations of novel identity constructions. Consequently, online dating ensures homogeneity in the types of relationships that are sought and found online.

.

"Anyone Can Do It": The Normalization of Internet Dating

It is impossible to understand how online dating has become mainstream without mentioning the rise of Internet technology. Access to the Internet rose steadily throughout the 1990s. In 1995, only 9 percent of adults in the United States were online. This figure increased to 56 percent by 1999 and to 67 percent by 2003 (Harris Poll 2003). Along with increased access to the Internet, its ease of use has increased as low-cost, high-speed connections have become available to more people. Also significant is the demographics of Internet users. Initially, the use of Internet technology was limited to the young, affluent, and highly educated. Though these groups are still slightly more likely than older, poorer, and less educated groups to use Internet technology, a recent Harris Poll (2003) reveals that the Internet population is beginning to look more like the general US population in terms of education, income, and age. As Internet use expands to include more people, it provides the mass audience needed for Internet dating.

When Internet technology was first made available to the general public in the mid-1990s, it was enthusiastically hailed as an innovation that would fundamentally alter the way individuals accomplish the routine tasks of life. Today, the Internet is used to do practically everything from reading the newspaper, paying bills, buying a home, searching for a job, taking educational courses, and purchasing consumer items. Finding a partner through the Internet represents just one more of the many activities that the technology enables.

In addition to technological innovation, demographic changes have contributed to the growth of Internet dating. According to social historian Barbara Dafoe Whitehead (2003), in the last three decades there has been a tremendous increase in the pool of people seeking mates. Thirty years ago, the dating pool was limited to young people who had never married. Today, it includes never-married men and women across a much wider age range, because both men and women marry at much later ages. In addition, the high divorce rate has created a large number of people who are looking for second and even third relationships. As well, older people who are living longer are also seeking companionship. Finally, the rise in the legitimacy of gay and lesbian relationships has propelled these individuals into the open market for relationships. Gays and lesbians are now able to seek partners through more conventional channels than they did thirty years ago, when they suffered greater public condemnation of their relationships. A large audience of actual and potential online daters has been created by these cultural changes.

While technological and demographic changes are part of the explanation for the rise of Internet dating, they are not sufficient to explain why it has become such an enormously popular and commonplace activity. Equally significant is the purposeful effort by marketeers to construct online dating as a legitimate way for ordinary people to meet partners. In order to increase revenues through paid customer subscriptions, marketeers of online dating sites have deployed strategies to increase their mass appeal (Sullivan 2002). One approach has been to promote online dating websites as places where romantic relationships are easily acquired by all participants, a strategy evident in print advertisements and on network and cable television.

Typically, advertisements for online dating services promise quickly to transform unhappy, lonely, single people into blissful, content couples, if they just take the initiative to join and post a personal profile. For example, the advertising copy of a recent television ad campaign by Match.com, an online dating site that claims to have the largest number of personal profiles, asks potential members: "Will you ever find the person who will change your life?" This rhetorical question is, of course, followed by an emphatic "Yes!" By using the free guide from Match.com entitled "How to find the right person in 90 days," the ad implies that finding a partner online is so easy that *anyone* can do it. Advertisements such as these present online dating as an efficient and utterly conventional activity. Moreover, they help convince the public that Internet dating is a viable way to meet a partner.

The legitimacy and appeal of online dating is further enhanced by the prominent suggestion that it is fundamentally about realizing the relationship goals of romance, love, and monogamous coupling. Regardless of whether online dating services are intended for heterosexual, gay, or lesbian users, they are typically constructed as places where conventionally established ideals of intimacy can be attained. For example, a visitor to the home page of EHarmony, a popular website for heterosexuals, is told that it is the website to use "when you are ready to find the love of your life." Similarly, Match.com proclaims itself to be "the world's number 1 place for love." Almost identical declarations are made on dating websites intended exclusively for lesbians and gays. Visitors to Planet Out will immediately notice the advertising copy that reads: "Find your Mr. or Ms. Right now!" Just underneath the bold headline is a link to the personal ads that reads: "Find Love." Although the home page of Gay-FriendFinder, an online dating site for gay men, carefully alludes to the possibility of a purely sexual relationship, it too makes love and romance a central part of its purpose. Its banner exclaims: "Find sexy single men for dating, romance, and *more*"[italics mine]. For additional emphasis, the headlines plastered across the home pages of heterosexual and gay and lesbian dating websites are routinely accompanied by visual graphics that conspicuously display stereotypical images of love and romance. Pictures that accompany print banners typically show two people holding hands, locked in a warm embrace, or gazing into each other's eyes over a candlelit dinner.

The inclusion of gays and lesbians within the rubric of love, romance, and monogamy is ironic precisely because gays and lesbians have historically been seen as incapable of achieving the relationship ideals typically linked to heterosexuality. Indeed, the idea that gays and lesbians are so far outside the boundaries of intimate convention that they cannot sustain intimate relationships based on love and monogamous commitment is an argument made by gay marriage opponents today. Of course, gay and lesbian patterns of intimacy do not necessarily preclude love, romance, and monogamy, but these options are not always the fundamental criteria around which lesbians and gays construct their relationships. A variety of historically unique types of intimate relationship characterize gay and lesbian subcultures: serial monogamy among lesbians, gay male subcultures based on sex, lesbian "Boston marriages" where physical intimacy is apparently absent, and butch–femme relationships that play on a heightened awareness of gender. Dating websites ignore these complex relationships in favor of assimilating gay and lesbian intimacy into a framework modeled on heterosexual standards of intimacy.

Indeed, most dating websites follow a similarly generic formula that includes the relentless depiction of words and images associated with heterosexual romance and a calculated muting of the sexual possibilities that might inspire or follow online encounters. Certainly, many online services exist primarily to link people who desire to meet others only for sexual activity, but websites whose business is limited to dating intentionally desexualize their content. The specific rules many dating sites have for creating profiles illustrate this point. For instance, Match.com expressly prohibits "overt solicitation for sex or descriptions of sexual activity, anatomy, etc." Similarly, YahooPersonals.com warns prospective ad writers not to "post detailed descriptions of physical characteristics or the types of sexual activities that interest you [sic them]" and forbids any video greetings that contain nudity or sexual language. Many dating sites also preempt the potential for any relationship that might develop outside the boundaries of monogamous coupling by forbidding the "solicitation of multiple or additional partners." In addition, some online dating sites seek to ensure that subscribers who deviate from the normative standards of heterosexual coupling are excluded from participation. People who are married, partnered, incarcerated, or under age eighteen are generally not permitted to post profiles.

Also significant is the absence of questions about sex on the lengthy questionnaires that prospective members of dating websites must fill out before they make their profiles available to other members. Most dating websites require the completion of a comprehensive questionnaire that covers minute details regarding the social, recreational, and relational interests of the applicant and those attributes sought in a partner. The questionnaires typically include inquiries about the kinds of sports activities members enjoy, the type of pets they have or would like to have (or not), the foods they eat, their sense of humor, their political views, and even their astrological sign. Questionnaires on dating sites exclusively targeted at gays and lesbians generally contain additional questions about identity disclosure (that is, how "out" the person is, questions about a person's membership in established lesbian and gay subcultures, for example "butch–femme," "lipstick lesbian," "leather," etc.). Each of these categories is usually covered in extensive detail. For example, the question on the Match.com questionnaire that asks "What kind of sport and exercise do you enjoy?" lists twenty-two activities ranging from aerobics to yoga. But curiously absent from the questionnaires are any questions about the type of erotic and sexual practices users enjoy and/or are seeking.

It would seem that people interested in finding someone to connect with, on an intimate level, might want to know something about the sexual desires, interests, and experiences of their prospective partner. Yet questionnaires on both heterosexual and gay and lesbian dating sites are entirely devoid of questions about prospective partners' definition of sex, the kind of sex members expect to have, the sexual experiences they have had, where they like to have sex, how often they like to have sex, or whether they even want to have sex. A few dating sites that serve gay men do ask whether the prospective member is specifically looking for sex, but remarkably the questionnaires on these sites also fail to ask any detailed questions about sexual practice preferences. The exclusion of explicit questions about sex on gay male questionnaires is particularly surprising in light of the fact that gay men have established subcultures of intimacy that are based entirely on sex. The omission of inquiries about sexual desires and interests serves to normalize the practice of online dating by cleansing it of the taint of sexual perversion.

Almost without exception, Internet dating sites are marketed to mass audiences as user-friendly venues where heterosexual, lesbian, and gay participants can secure the relationship goals historically associated with idealized heterosexuality: namely, a long-term, monogamous, and preferably connubial relationship between two people. This vision is reinforced by the "success" stories regularly found on the homepages of dating websites. Couples who have realized a committed and exclusive relationship, or become engaged or married as a result of "meeting" through a particular online dating service, are counted among the website's success stories. To reinforce this point, online dating websites routinely publish the sometimes lengthy testimonials—along with photos of course—of members who have secured their relationship through the site. Typical are narratives that make reference to the esteemed status of couplehood or that invoke the idealized concepts of "romance" and "soul-mate":

> Gay FriendFinder helped me find my soul mate. I work hard and don't really have time or energy to go out to bars and clubs to meet people. I tried Gay FriendFinder mostly out of curiosity and met Jeff—he's too good to be true! Thank you, thank you, thank you (BizGal28).

> I just wanted to tell you that I have found that special someone and also wanted to say thanks. If it hadn't been for your site [curiouslove.com], I would probably be single and very miserable. Just wanted to say that your site is very awesome …

> As Ryan walked me to my car we kissed again. And he invited me to spend the following day wandering in San Francisco. I had really planned to spend the day getting a bunch of errands done, but his talk of sipping tea at the Japanese Tea Garden and a romantic picnic in Golden Gate Park was too much to pass up. We decided to meet the next morning. For the first time in years I've found a man that I can have a real relationship with (Julie; perfectmatch. com).

On heterosexual dating websites, testimonial narratives are often supplemented by statistics on how many marriages have been produced through the site. An illustration of this common practice can be seen on EHarmony's home page. Here, a selection of smiling, apparently happy, hand-holding couples is continuously flashed on the background of the page along with the prominent display of the date of their first meeting and the date of their subsequent engagement or marriage. With the exception of marriage, the definition of "success" in online dating does not appear to vary by sexual orientation. A successful online dater is one who has secured the ideal type of pairing historically linked to heterosexuality—a monogamously committed coupling of two people who are thereafter forever linked through romance and love.

The visual images and advertising copy displayed by online dating services make it clear to potential subscribers that these are not venues in which to find casual sexual encounters, non-monogamous relationships, or experiment with new gender or sexual identities. They are places where one can safely seek and find intimate relationships that embody the ideals of love and romance. Consistent reinforcement of the idea that romance, love, and long-term monogamous coupling can be realized through online dating eases the public's fear that sexual deviants are lurking behind the online profiles. Moreover, it induces confidence in the belief that placing or responding to an advertisement online is not a stigmatized activity undertaken by sexual deviants and losers, but an activity that anyone can easily and safely engage in.

Technological and demographic changes, along with deliberate marketing strategies that link Internet dating with conventional relationship goals, have helped make the process of seeking and finding partners through online personal advertisements attractive to a mass audience. Today, people who, in the past, may have hesitated to meet someone in a virtual space eagerly participate in placing and reading online personal profiles. Indeed, Whitehead suggests that Internet dating is "likely to be as influential in shaping the patterns of mating in the early 21st century as the internal combustion engine was in shaping patterns of youthful dating in the early 20th century" (2003: 175). If her analogy is accurate, it is useful to consider some of the possible consequences of online dating.

Expressway to Romance: The Increased Efficiencies of Online Dating

One of the more obvious effects of online dating is that it increases temporal efficiency in the search for a partner. Traditional methods of finding a partner require intense investments in time. One must first find a compatible mate through workplace, school, or family networks, and then spend time with the prospective person to discover whether there is potential for pursuing the relationship. But online dating allows users to establish the specific qualities they are seeking in a partner prior to meeting him/her. To ensure that a prospective partner has the desired characteristics, users of online dating websites may select the age, race, and sexual orientation of the person they are looking for; read the biographical profile of a prospective partner; scan photos; and, if available, watch a video or listen to a voice greeting. In short, the online dater can ensure that a prospective date meets all of the requirements for a relationship prior to any actual contact with the person. Online dating effectively functions as a labor-saving device in the search for a partner.

Facilitating the efficiency of online personals are new and increasingly sophisticated software programs that allow users to select partners who are likely to have compatible personality traits. Several online dating websites have hired social scientists to develop tests that purport to measure the personality and character traits of their members. Though the validity of these tests is suspect, they are used to increase the likelihood that people will find a well-suited match. For example, EHarmony boasts that its "personality profile" measures twenty-nine dimensions of personality that are scientifically proven to predict long-lasting relationships. Perfect Match offers a similar test that measures a member's character traits and value orientation to better fit him/her with someone who shares a common "relationship style." Presumably, these tests make the process of partner-seeking more efficient because they eliminate the possibility of subjective "error" in selection.

Online dating not only increases the efficiency of partner selection, but also expands opportunities for people to find partners. Because they draw from a national (and even a global) pool of applicants, online dating sites provide a far larger pool of potential eligible partners than conventional social networks typically allow (Whitehead 2003). As such, online dating is particularly useful for members of sexual and/or racial/ethnic minorities who are typically limited in their search by a small pool of eligible partners. Simply stated, online dating websites that serve specific identity groups make it easier to locate partners. The many dating websites hosted in the United States intended specifically for Blacks, Latinos, Asians, transgendered, and gay people, allow them to reach an audience outside the confines of their geographically limited social networks.

In theory, online dating should also expand opportunities to create new forms of relationships, courtship patterns, and identity expressions. In the online environment, relationships can take place entirely in virtual space. Contact between a reader and the originator of an online profile may begin in a public chatroom and then proceed to private emails and perhaps Instant Messages, but need never be realized in a face-to-face encounter. In this sense, online dating provides the potential to go beyond existing categories of gender, sexuality, and even race, because relationships that take place entirely online are not mediated by voices, bodies, smells, or—in the case where pictures are unavailable—faces. New forms of sexual relationships may also be defined because online sexual partnerships can and do develop without physical sexual contact. As well, traditional patterns of courtship where men have historically been the initiators may also give way to greater freedom for women to initiate romantic and sexual encounters and exercise control over the process and content of their interactions. But the realization of these possibilities is incomplete, largely because online dating websites construct intimate relationships along the constricted confines of romance, love and monogamy, and rely on existing categories of sexuality and gender to make the sale.

Consuming Love: The Commercialization of Intimacy

Online dating transforms the search for intimate partners into a consumer activity. The process can be likened to a retail shopping experience that provides patrons with expansive options in partner selection. Each dating website resembles a store that stocks an enormous variety of "products." Shoppers who visit an online dating website browse among the many items available and, like their counterparts in the mall, specify the size, color, and overall quality of the one they are seeking. If buyers don't like what is offered, they can easily move on to the next store until they find exactly what they are interested in. Moreover, if the product does not perform as promised, shoppers may return to the store for a replacement model. Similar to consumer protection agencies that police conventional retail shops, websites like Truedater.com allow online daters to "turn in" writers of advertisements who are less than candid about their appearance, or marital, or financial status. Indeed, online dating transforms people into rational consumers who scrutinize the marketplace for the "best available deal" on intimate relationships.

A consumer market model may provide the greatest number of choices for people who are looking for a date, but it simultaneously reproduces the boundaries of existing gender and sexual identity categories and, therefore, may actually limit the relationship choices people have. Like their more conventional retail counterparts, online dating websites categorize the products they sell in a way that makes them easy to find. As such, they structure the options shoppers may select along the lines of established identity categories that consumers easily recognize. When shoppers search for partners online, they make their initial selection on the basis of gender, sexual orientation, race, and age. Already embedded in the software are the categories "man," "woman," "gay," "straight," "white" and "black." Online daters decide on the prescribed criteria for the "product" they want and then the computerized sorting mechanism returns only profiles of those people who represent the specific categories selected. The online format does not permit people to consider or define alternatives to the categories already given.

Gender, sexual orientation, race, and age are invisible in virtual space; therefore, online dating contains the potential to create relationships that are modeled outside the boundaries of these established identities. Online dating websites could, for example, allow users to define searches around specific personal character traits, shared interests, or life goals. To be sure, these criteria are often used in the secondary aspects of a partner search, but the primary step in the selection process involves choosing candidates by gender, sexual orientation, age, and race. The paradox here is that the very efficiencies of online dating that expand the possibilities of finding partners also confine the parameters of the search and, therefore, limit the prospects of expanding conventional constructions of intimate relationships.

The potential for online dating to transcend established identity categories is further constrained by the fact that, as mentioned earlier, online dating sites are increasingly targeted toward particular sexual orientation, racial/ethnic, social class, religious, and age groups. Inarguably, the separation of dating websites by specific identity categories makes it easier to find someone who meets one's desires. After all, gay people don't ordinarily yearn to date heterosexuals. But the construction of online dating sites by narrowly-bounded definitions of gender and sexuality compels the users of online websites to express their allegiance to a set of fixed identities prior to engaging in the activity of searching for an intimate partner. Online dating websites do not encourage users to explore the space in between categories, nor do they promote the possibility of creating new ones beyond existing constructs.

Because online dating websites compel users to identify with established gender and sexual categories, they may also encourage writers of personal profiles to rearticulate rather than transform the boundaries of gender. Though social scientists have yet to produce a systematic study that explores gender expression in online advertisements, a casual overview of online profiles suggests that writers typically adhere to established social meanings around masculinity and femininity. For instance, the online profiles of heterosexual and lesbian females routinely make references to a desire for love and romance. In contrast, both gay and straight male profiles tend to describe the physical attributes of the partner(s) they are interested in. If online personal profiles solidify rather than expand conventionally understood meanings of gender, they too are unlikely to offer possibilities for creating relationships outside of conventionally established frameworks.

Online dating represents a spectacular change in the process of finding partners, and provides more efficient ways for people to meet their relationship needs, but it has yet to transform prevailing ideas about intimate relationships. Just as Internet users shop the net for retail items, they can search the global marketplace for intimacy. Indeed, finding relationships in virtual space now has mass appeal. But online dating websites sustain their mass appeal through the insistent and ever-present reliance on a particular relationship model that embodies the characteristics historically tied to heterosexual couplings. Online dating websites construct ideal relationships within the boundaries of convention. The trilogy of romance, love, and monogamy dominates the online dating scene, while alternative models of intimacy and the sexual possibilities of intimate relationships are de-emphasized. Moreover, online dating websites provide few opportunities to contest socially imposed boundaries around sexuality and gender because these recognized identity categories are embedded in the very structure of the websites themselves. As a result, Internet dating strengthens rather than expands the boundaries of the categories through which people imagine their intimate relationships and, therefore, limits ideas about alternative forms they might take.

Note

1. The writer thanks Susan Bass and Anne Colvin for helpful comments on earlier versions of this essay.

Section 4

Mobility

Domesticating New Media

A Discussion on Locating Mobile Media

By Larissa Hjorth

Introduction

As convergence leaves its mark as this century, the ultimate alibi in the convergence rhetoric seems to be the mobile device. Convergence can occur across various levels such as technological, economic, industrial, and cultural. As Henry Jenkins observed, in the growth of mobile phone into converging various forms of multimedia—into the ambiguous and yet ubiquitous mobile media—one could almost forget that mobile media arose from an extension of the landline telephony.[1]

Now the twenty-first century's equivalent to the Swiss army knife,[2] mobile media encompasses multiple forms of media including camera, gaming platform, MP3 player, and Internet portal. As we begin to chart the burgeoning phenomenon of mobile media, we must reassess the methodologies and frameworks being used. How do we grapple with mobile media's interdisciplinary background? Should mobile media be framed in terms of the mobile communication and material cultures traditions, fathered by British theorist Roger Silverstone, that have contextualized the sociocultural processes of media technologies in terms of the domestic technologies approach? Or should mobile media be framed by creative theories and practices of new media?

The rise of the mobile phone into mobile media has attracted scholars from various disciplines such as media studies, gender studies, cultural studies, media sociology, virtual ethnography, and new media, all bringing with them a wealth of traditions, methodologies, and approaches. One of the dominant and highly successful approaches in the field of studying mobile phone cultures is, undoubtedly, the domestic technologies approach.

As an interdisciplinary framework, the domestic technologies approach[3] draws from anthropology,[4] cultural studies,[5] and consumption studies.[6] A significant part of its lineage lies in anthropology and its commitment to analyzing the processes of material cultures in everyday life. Undoubtedly, the seduction of the domestic technologies approach is that it focuses on the symbolic dimensions of technologies in everyday life. In particular, the domestic technologies approach focuses on meanings individuals and cultural contexts give to their technologies, extrapolating on the ways in which users perceive them.

However, as the mobile phone expands into a multimedia device, how can the dimensions of social and reproductive labor—addressed by domestic technologies approaches—be incorporated into the growing realm of mobile media as new media? Domestic technologies approaches seem to fail in grasping the role of creative labor associated with mobile media beyond social and reproductive labor paradigms. In turn, new media approaches to mobile media seem unequipped to address the political dimensions of social and reproductive labor. Since both approaches have been useful in addressing the dynamic, social, creative, and procedural nature of mobile media, it seems fitting to discuss these two enveloping traditions in the context of locating mobile media within "Domesticating new media."

In this [selection] I will explore the marriage between the two traditions—on the one hand, the domestic technologies approach, on the other hand, new media remediation approach—in order to conceptualize some of the paradoxes found in mobile media in terms of earlier, ongoing processes. I will outline some of the key attributes and paradoxes that have plagued both traditions' examination of mobile media. Through the example of mobile location-aware gaming, I will draw upon current discourses around mobile media and its coinhabitation in both domestic technologies and new media discourses. As this [selection] will argue, through mobile media we can gain insight into some of the recurring paradoxes that run across disciplines and boundaries, continuing to haunt and limit interdisciplinary approaches to twenty-first-century new media practices. In particular, I argue that the emphasis upon visuality and screen-centric views have neglected to address one of the most important aspects of mobile media, the haptic.

Media @ Mobile

Mobile media is a strange animal to tame. Part domestic technology, part new media, the phenomenon has attached much stargazing and posturing about the future. Through the portal of mobile media, we have witnessed mobility becoming conflated with futurism. The rise of the mobile phone has been marked by its shifting symbolism, usages, and adaptations.[7] When mobile phones first graced the mainstream in the 1980s they were associated with yuppies and conspicuous displays of wealth as demonstrated in the iconic 1980s film *Wall Street*. Then, as mobile phones were adopted and adapted by youth cultures, the phone shrunk into a complex creature adorned by user-created customization from phone straps to sticker faceplates and screen savers. Then, as the phone became more than *just* a phone and started to emanate this century's Swiss army knife, it expanded in size both physically and psychologically to become an integral component in visual, textual, and aural practices in contemporary everyday life.[8]

It is with this size change that we moved into an epoch of mobile multi-modality that became synonymous with contemporary mobility. The rise of mobile media as multimedia par excellence has also been accompanied by corporate smoke and mirrors around the so-called empowered user by way of user-created content (UCC) and pro-sumer agency. In this climate of optimistic futurism, mobile media promised a further democratization of media. But as Finnish theorist Ilpo Koskinen notes, this accessibility of multimedia often resulted in the aesthetics of banality; images and media rehearse well-known genres and themes.[9] Within the so-called banality are normalized power relations inscribed at the level of everyday practice; thus mobile media serves to remind us of the growing significance of place.

Much work has been conducted around the "banality" of mobile media practices in terms of cameraphone visual and distribution characteristics with many theorists pointing to the content of mobile media rehearsing earlier media (that is, camera-phone images reenacting analogue genres) being banal but the context in which they are shared (or not) providing much signification;[10] however, it seems that the haptic economies, so particular to mobile media, are in need of reevaluation. While Mizuko Ito and Daisuke Okabe's 3 S's—sharing, storing, and saving[11]—noted some of the particulars, we need to examine the politics of "waiting for immediacy" just outside the frame/screen. In other words, what are some of the haptic workouts occurring just outside the frame that undoubtedly affect inside the frame?

So what do I mean by haptic? Just one glance at the current models of mobile media such as iPhone and LG prada and we can see that the screen is no longer about visuality; it is about haptics—haptic screens, to be precise. The engagement of mobile media is not ocular in the case of the gaze or the glance, but rather akin to what Chris Chesher characterizes as the "glaze."[12] Drawing on console games cultures, Chesher identifies three types of glaze spaces—the glazed-over, sticky, and identity-reflective. For Chesher, these three 'dimensions' of the glaze move beyond a visual economy, deploying the filters of the other senses such as aural and haptic.

The haptic has often been undertheorized in mobile communication discourses, often left up to new media practitioners to grapple with in such projects as location-aware gaming. In the growth of mobile-media discourses, much has been discussed in terms of media such as cameraphone practices and the associated sharing and distribution methods. However, much of the rhetoric around mobile media and convergence has been focused upon the frame and visuality—as such concepts as "cross-platforming" entail. These models have discussed media in terms of twentieth-century preoccupations with the visual and the screen, neglecting to reorient frameworks around what makes mobile media so particular; whether being mobile or immobile, the logic is of the haptic. It is about the touch of the device, the intimacy of the object, that makes it so meaningful.

For new media artists such as Rafael Lazano Hemmer and his relational architecture projects, it is this very oscillation of the haptic and the cerebral that partakes in mobile media copresence that makes it such a particular vehicle for twenty-first-century new media practice. In urban spaces, it is not so much the cameraphone images that are transforming the spaces but, rather, the haptic workouts of the everyday user documenting. Much of the discussion of mobile media has encircled the important role of mobile media copresence,[13] and yet the integral notion of the haptic, apart from the hype around SMS thumb cultures, has been largely ignored.

However, the critique of normalized everyday practices and the haptic workouts outside the frame can be found in the various upsurge of experimental new media projects such as location-aware gaming, mobile gaming, or "big games." Location-aware or pervasive games often involve the use of GPS (geographic positioning systems), which allows games to be played simultaneously online and offline. As Finnish theorist (and director of DiGRA) Frans Mäyrä notes, gaming has always involved place and mobility and yet this is precisely what is missing in current games, especially single player genres.[14] Mäyrä points to the possibilities of pervasive (location-aware) gaming as not only testing our imagination and creativity but also questioning our ideas of what constitutes reality and what it means to be copresent and virtual.

The notion of "big games" does not so much relate to the gadget's gluttonous size but rather it has more to do with the role of people and the gravity of place in the navigation of copresence. These projects served to remind us of the importance of locality and its relationship to practices of copresence. The potentiality of "big games" to expose and comment on the politics of copresence—traversing virtual and actual, here and there—in contemporary media cultures has gained much attention. They highlight some of the key paradoxes of everyday life that have been exemplified in mobile-media projects such as location-aware gaming. The paradoxes include virtual and actual, online and offline, cerebral and haptic, delay and immediacy.

As Frank Lantz, a New York-based game designer who has been involved in such pivotal projects as *PacManhattan*, notes, the importance of location-aware mobile gaming—or "big games"—definitely plays an important role in the future of gaming.[15] Citing examples such as *PacManhattan*, United Kingdom's blast theory, Geocaching, and Mogi, Lantz emphasizes the importance of these projects in testing the notion of reality as mediation. As Lantz observes, the precursors to big games and the 1970s New Games Movement were undoubtedly the art movements of the 1960s such as happenings (impromptu art events) and the Situationist International (SI) tactics of Guy Debord such as *detourement*, which operated to interrupt/disrupt everyday practices and the increasing role of media and commodification. In this way, this can be paralleled with the trend in contemporary art from 1990s that French curator and critic Nicolas Bourriaud dubs "relational aesthetics."[16] As Bourriaud observed, "relational aesthetics" dominated the international art scene from the 1990s onwards, building from an emphasis upon locality and deinstitutionalization of installation and the "international" in favor of the vernacular and local.

Locative mobile gaming illustrates the paradoxes of mobile media as part of the cyclic and dynamic processes of technology. For example, in an age of so-called immediate technologies, such projects enlighten us to the conundrum of instantaneity, that is, the inevitable poetics of delay. They highlight the price of mobility and its oscillation between freedom and leash[17] in which work and leisure boundaries are increasingly blurred.[18]

Locative mobile gaming also emphasizes other paradoxes apart from the aforementioned immediacy/delay temporal conundrum. These projects highlight the way in which mobile media can often interfere with, rather than help, face-to-face connections. For example, the tyranny of mobile media's creative labor/democratizing of media dimensions, as epitomized by UCC, sees users becoming more enslaved to the technology rather than it freeing up time to spend with intimates. Locative mobile gaming projects afford us one way in which to reflect and mediate on the paradoxes of contemporary mobile media.

Moreover, locative mobile gaming illustrates that in the face of democratizing of media, new media is still far from the understandings and interests of the everyday person. It also reflects new media artists' fears and yet curiosity about mobile media's ultimate creative conundrum: is it the rise of democratized media and mainstreaming of new media or does the "banality" represent the domination of pedestrianization of new media? Can mobile media teach new media ways to remember the histories "shock of new" as actually the "delay of the banal"?

As a new conflation of many techniques, traditions, and media histories, it is no easy task to outline the nebulous terrain of mobile media. In this [selection] I argue that one way in which we can understand mobile media is vis-a-vis its borrowing from, and adapting of, various sociological and new media traditions. In the next section of the [selection] I will address two traditions—domestic technologies and remediation, new media approaches—in the cartography of mobile media. I argue that many parallels can be found in the two traditions and that by incorporating the two genealogies we could gain much insight into mobile media.

Just as mobile media needs the rigor of domestic technologies approaches to comprehend the social dimensions of new media, it also needs the innovative approaches of new media theory in order to reconceptualize the conflations between creative and social labor in mobile media's fusion between media communication and new media practices. Through the conflation of domestic technologies and remediation of new media approaches we can begin to conceptualize mobile media as no longer just a "third screen," but, more importantly, a *third space*.

Domesticating New Media: Two Examples of the Multitraditions of Mobile Media

The rise of mobile media could be read as nascent. However, such a belief, propagated in global media's lauding of the new mobile revolution in consumer agency (in the form of the prosumer and Web 2.0), neglects to address the dynamic dimensions of technology as a sociotechnological process. In the case of domestic technologies approaches, in which domestication is always an ongoing and never-completed process, the dynamics of mobile media extends already existing cyclical models. So too, in the tradition of new media, in which old and new have had a dialectical and dynamic relationship that disrupts any linear or casual notion of time.

As mobile communication and media industries converge, the all-pervasive futurist rhetoric becomes stifling. And yet, if the twin histories of new media and mobile communication have taught us anything, the "new" is always remediated and mediated. Each "new" technology deploys techniques of the older technology, which in turn revises the earlier media. This cuts to the core of all communication and cultural practices implicated in intimacy. For Jay Bolter and Richard Grusin, new media are remediated with older media into a dynamic ongoing process that disrupts any causal or linear notion of old and new technologies.[19] As Margaret Morse concisely notes in the case of the Internet, all forms of intimacy are mediated—by language, gestures, and memories.[20] Emerging forms of visual, textual, and haptic mobile genres such as SMS and cameraphone practices—reenacting earlier rituals such as nineteenth-century letter writing, postcards,[21] and gift-giving customs[22]—have only served to highlight the remediated nature of the rise of mobile media.

There is much to be learnt from understanding the parallels between new media theory on remediation and mobile communication's usage of the domestic technologies approach. Like the domestic technologies approach,[23] the study of new media through the lens of remediation echoes a similar philosophical stance. As influential theorist in the field of media archaeology, Erkki Huhtamo, has argued, the cyclical phenomena of media tend to transcend historical contexts, often placating a process of paradoxical reenactment and reenchantment with what is deemed as "new."[24] For Huhtamo, media archaeology approaches are "a way of studying recurring cyclical phenomena that (re)appear and disappear over and over again in media history, somehow seeming to transcend specific historical contexts."[25] As Jussi Parikka and Jaakko Suominen note, the procedural nature of media archaeology approaches means "new media is always situated within continuous histories of media production, distribution and usage—as part of a longer duration of experience."[26]

Citing an example of the launch of Nintendo DS that heralded a new and "unique" experience for twenty-first-century entertainment, Parikka and Suominen note that much of contemporary postindustrial digital media culture is inundated by futurism that seeks to break with the past.[27] Parikka and Suominen note that this "creates the impression that, in the new media discourse, the past functions solely as something worse or less sophisticated, something that has to be left behind and practically forgotten."[28]

When Marshall McLuhan identified that the content of new media is that of the previous technology, he highlighted the nonlinear and dynamic role of new technology imbued by the specters of old technology.[29] In short, that the "new" is far from superseding or breaking with the old as modernist mythologies would have it. The fact that the notion of "new" in new media has been continuously challenged and demonstrated as a fallacy echoes the way in that technology has been approached by many mobile communication scholars (from predominantly sociological and urban anthropological traditions) through the domestic technologies approach.[30]

In the picture painted by the domestic technologies approach, domestic technologies such as the radio, TV, and mobile phone are seen as part of the cyclic and ongoing process of consumption in everyday contemporary life. As Daniel Miller notes in his coauthored study with Heather Horst on Jamaican cell-phone use, "what one has to study are not things or people but processes."[31] Much of the literature analyzing mobile communication has utilized the domestic technologies approach[32] to identify adoption and adaptation of technologies as always ongoing and never completed.[33] Like the cultures in which they inhabit, domestic technologies are always in flux. Domestication is ongoing and dynamic, and through customization practices we can domesticate domestic technologies as much as they domesticate us in a productive tension. In the case of the mobile phone, while the domestic technology device may have *physically* left the home, it *psychologically* resonates what it means to be at home and local no matter where it is located.

As David Morley has noted, the mobile phone has often been cited as a key example of domestic technologies par excellence.[34] Key scholars in this area include Leslie Haddon, Roger Silverstone, Rich Ling, and Miller. As a key scholar in the field, Haddon provided decisive apparatus to comprehend the dynamic and enduring processes of the domestic technologies.[35] Often users' relationships to their domestic technologies can wax and wane, drawing feelings of ambivalence, and yet inevitably due to the prescribed need to have the technologies to be part of contemporary urbanity.

As an approach, the domestic technologies method sees the process of engagement with technologies undergoing various stages or nodes of a cycle that include "imagination, appropriation, objectification, incorporation, and conversion."[36] Consumption is seen as an ongoing process that is perpetually negotiated, way after the actual point of sales. As Rich Ling notes, "our consumption becomes a part of our own social identity. Further, others' consumption is a type of lens through which we see them and through which we interpret their social position."[37]

Mobile media represents a meeting of the crossroads between the genealogy of domestic technologies and media archaeologies of new media. In both these traditions, we see mobile media remediating and reenacting previous media cultures and modes of domestic regimes. Reminding us of our forgetting whilst harnessing the inevitable amnesia that accompanies any notion of "new," mobile media represents the conundrum of new technologies. In new media discourses we can find many examples of the content or specters of the older media. Like the domestic technologies approach, the study of new media through the lens of remediation echoes a similar philosophical stance.

According to Timo Kopomaa, the mobile phone is an extension of nineteenth-century media.[38] For Kopomaa, mobile media creates a new "third" space in between public and private space. On the one hand, the project of examining mobile media entails observing the remediated nature of new technologies and thus conceptualizing them in terms of media archaeologies.[39] On the other hand, mobile media's reenactment of earlier technologies is indicative of its domestic technologies tradition that extends and rehearses the processes of precursors such as radio and TV. It is the fact that Kopomaa draws our attention to mobile media as a third space, rather than third screen, which is significant.

Both traditions—the domestic technologies and new media remediation approaches—emphasize the cyclic and dynamic process of media technologies that cannot be simplistically divided between old and new or inside and outside the screen. Rather, the cartography of mobile media is one imbued by paradoxes. In the case of cameraphone practices—whether still or moving—mobile media demonstrates two distinctive paradoxes, that of the *reel* in the real, and the inherent poetics of *delay* in the practice of immediacy in the navigating of offline and online copresence. As Lev Manovich identified,[40] contemporary new media and digital practice are all consumed by fetishizing the real through the lens of the reel—that is, texture and skin of the analogue.

For Manovich, the way in which to understand the remediated emerging digital cultures and the haunting by the ghost of the analogue is through a series of paradoxes. These sets of paradoxes are located around the relationship between the real and the reel. As Manovich identifies, while the analogue may disappear, it will continue to haunt the digital in the form of the analogue's particular realism, the "reel." This is evident in the way in which cameraphone practices echo previous analogue norms[41] and that, in turn, make mobile media, according to Koskinen, characterized by "banality."

However, one of the most compelling examples of the real/reel phenomenon, where the tactile process of the analogue is fully felt both metaphorically and actually, is the rise of screen cultures in mobile media. In particular, the rise of such mobile-media devices as iPhone, LG prada, and Samsung Arami phone—to name a few—all incorporate one key feature, haptic screens. Here the reel/real paradox is played out in the haptic versus visual, in which the haptic is undoubtedly the more meaningful factor that "domesticates" the device into the user's everyday life. Much of the specters of the analogue reel are more about the tactical experience of image processing; and while these processes have been deleted in the rise of the digital, it is the legacy of the haptic—that has moved from the filmic developing process to the actual politics of the touch screen—that continues unabated. However, in the language set of twentieth-century media cultures, much discussion was given to visuality rather than the increasing role of the haptic.

While location-aware projects are invaluable in geocaching (such as GPS) and demonstrating the importance of place and specificity in a period of global technologies, they also served to highlight one of the greatest residual paradoxes of mobile media as a metaphor for sociotechnologies. That is, the paradoxical politics of copresence. One example can be found in the aims of twentieth-century technology to overcome difference and distance from geographic and physical to cultural and psychological. This attempt to overcome distance and difference sees the opposite result, the overcoming of closeness. Practices of copresence intimacies become fetishized, through what Misa Matsuda has characterized as "full-time intimacy."[42] This recites what Michael Arnold identified as the Janus-faced nature of mobile media that operates to push and pull us, setting us free to roam and yet attaches us to a perpetual leash.[43]

As Arnold notes, the Janis-faced phenomenon is symptomatic of what Martin Heidegger characterized as "un-distance." The role of technology in the twentieth century has always been to overcome some form of distance—whether geographic, physical, social, cultural, temporal, or spatial. But herein lies the paradox. The more we try to overcome distance, the more we overcome closeness. This is the kernel of un-distance and its temporal and spatial tenor. Un-distance can be seen today in the practice of mobile media, particularly pervasive location-aware projects that rely on the so-called immediacy or instantaneity of the networked.

However, one could argue that un-distance has been perpetuated by the ocular-centrism of twentieth century "tele" media, a phenomenon that has been disrupted by mobile media's emphasis on the haptic. For Ingrid Richardson, mobile media needs to harness the importance of the haptic. Conducting a small ethno-phenomenological study on the use of phone-game hybrids, Richardson disavows the ocular-centrism prevalent in "new media screen technologies" to focus on "the spatial, perceptual and ontic effects of mobile devices as nascent new media forms."[44] As she persuasively observes,

> In order to grasp the epistemic, ontic, and phenomenological status of screen media it is important to trace their ocularcentric legacy; by understanding this history we can then interpret how mobile screens in particular work to bewilder classical notions of visual perception, agency and knowing.[45]

Indeed, one of the compelling factors to arise from mobile media, and this links back to its fusion of remediation and domestic genealogies, is the persistence of the ontology of the reel. However, unlike the twentieth-century "reel"—in the form of the aural modes of address embroiled in "screen-ness"—the mobile reel, and thus possible creative worlds and realities, is undoubtedly governed by the haptic.

The game of mobile media—whether it be partaking in cameraphone imagery and the haptic exercises outside the screen, to mobile gaming, in which interactivity and engagement are navigated by haptic mobility and immobility rather than visualities of screen cultures—is undoubtedly changing how we are thinking about domestic technologies and new media. Through the lens of paradoxes that encompass virtual and actual, online and offline, haptic and visual, and delay and immediacy, some lessons about twentieth-century media practice can be learnt. For anyone that has participated in a mobile pervasive game, they will quickly identify the lack of coherence between online and offline copresence. The more we try to partake in the *politics of immediacy*, the more we succumb to the *poetics of delay*. This paradox extends beyond just mobile gaming and can be found in many of the multimedia possibilities of mobile media— from cameraphone imagery and MMS to moblogging and SMS.

In the case of the growing interest in urban screen cultures as an analogy for the twenty-first century, one could argue that it is indeed the very eruption of the twentieth century's obsession with visualities for the twenty-first century's politics of the haptic that dominates the canvas of mobile media. From the haptic screen interfaces to the various multimedia tactics such as mobile gaming that disavow the screen for the haptic and audio, mobile media revises the perceived hierarchies of the senses which, in turn, could breath new life into new media and domestic technologies approaches.

Time After Time: The Never-Ending Concluding Beginning

In modernism, the role of originality was celebrated. For the modernist avant-gardists, such vehicles as technology served as a decisive break from the past in what art critic Robert Hughes characterized as "the shock of the new."[46] In contemporary postindustrial digital cultures, the "new" promised by mobile media is in fact "banal" and located in nostalgic politics such as the real/reel paradigm. In this [selection], I have assembled two traditions—domestic technologies and remediation—in order to show the similar cyclic debates operating across disciplines. I argue that perhaps mobile media needs to be conceptualized as the "shock of the banal," that is, its paradoxes—online and offline, virtual and actual, delay and immediacy, haptic and visual—are far from new and can be traced through various disciplinary traditions.

In my ethnographic studies into cameraphone practices in Seoul, Tokyo, Hong Kong, and Melbourne, one of the increasing features of the tyranny of the "full-time intimacy"[47] of mobile media customization is the use of immediacy to camouflage delay. Many respondents spoke of creating their own forms of delay so that they could savor the SMS or MMS. With the immediacy of such technologies, the tactics of pretending to not see or receive a message immediately allowed respondents time—what I call "the poetics of delay." Moreover, the persistence of the reel in much of the cameraphone images, genres, and mobile movies was significant. In particular, for many respondents, mobile-media making was less about visual economies and more about aural and haptic modes of address akin to earlier "reel" domestic technologies such as the TV and radio. Thus, the conflation between domestic technologies and new media approaches could further address one of the greatest paradoxes of shifts from twentieth to twenty-first century media; that is, rather than it being a history predicated on visualities as the "screenness" would entail, it is a history of the rise of the audio and the haptic that are becoming the key indicators and characteristics of mobile media.[48]

I have chosen to focus on two traditions—one draws from media and communication and material cultures (anthropology) in the form of the domestic technologies approach; the other calls upon new media approaches of remediation and media archaeologies approaches. In these two traditions we can see various similarities—the focus on the dynamic, sociocultural processes of mobile media. While the former allows for more insight into social and reproductive labor debates, the latter affords us acuity into shifting modes of accessing creative labor and everyday life. In the case of mobile media projects such as location-aware gaming, I argue we need to draw upon the two models, incorporating them into a new framework for evaluating dimensions of mobile media, and twenty-first century screen cultures, in terms of key attributes such as the haptic. The important factor here is against the seductive and simplistic futurism prevailing in much discussion around mobile media we need to recognize that mobile media, like new media, is inevitably involved in the politics of the banal and nostalgia.

Much of the futurist posturing accompanying mobile-media discussions in global media have celebrated the potential democratization of new media. With the rise of the prosumer from the term coined by Alvin Toffler in 1980[49] to its adaptation by Don Tapscott in 1995[50] to the context of the Internet's digital economy, much of the media of late has celebrated UCC and the prosumer as part of the Web 2.0 enterprises. But behind this rhetoric is the pivotal role mobile media has played in creating and reenacting debates about technology, labor, and creativity that have long accompanied new media and domestic technology discourses. Rather than just domestic and artistic labor having little or no remuneration in the general community, now the UCC associated with mobile media could see the everyday person subject to the injustices of industry convergence whereby corporations buy and exploit social and creative labor in the form of Web 2.0 media such as SNS.

The conundrum of new mobile technologies is that they are *supposed* to free us up and yet, as a good existential crisis would have it, the freedom is a leash. Work becomes mobile; labor is on a perpetual drip. We are supposed to be available at all times, perpetually connected. Rather than free us, the "immediacy" logic of mobile technologies makes us feel like we must be quicker and must achieve more. Rather than saving time, applications such as cameraphone image making—and the attendant customizing and modes of sharing/distribution—mean users spend a lot of time sharing and editing the so-called immediate. The present gets put on hold. However, one of the features that becomes apparent in mobile media is the need to move beyond the screen-centric and ocularcentricism of twentieth-century media and reconnect with the very reason the mobile phone has grown into mobile media, its importance at the level of everyday haptics.

By engaging in the significance of the haptic in mobile media we can grasp some of the paradoxes at play. It is important to recognize that this conundrum of delay and immediacy is not new with the rise of mobile media. Rather, these paradoxes have been central in the emphasis upon screen cultures in the face of the importance of the haptic in the making sense of mobile media. As I have attempted to discuss, mobile media represents some interesting paradoxes about contemporary media and consumer cultures. In this [selection] I have tried to show the ambivalences surrounding mobile media from both a new media and domestic technologies approach in order to reconceptualize the philosophical and phenomenological dimensions of mobile media. To socialize the creative media dimensions and to innovate the social, domestic dimensions.

In order to grapple with the burgeoning field of mobile media we need to comprehend the twin histories—such as the domestic technologies and remediation approach—to fully grasp the histories, contemporary and future paradoxical permutations of mobile media and not just fetishize the "new" by futurist posturings. Mobile media is undoubtedly a project involving the domesticating of new media in which old boundaries between art and life, production and consumption perpetually change and shift, repeat and pause. But it is time that we moved away from the twentieth-century preoccupations with visual cultures and screen-ness that deems to view mobile media as a (advertising) *third screen* and instead acknowledge its genealogy as a *third space* that is governed by the politics and aesthetics of haptics.

Notes

1. Henry Jenkins, "Welcome to Convergence Culture," *Receiver*, 12 (2005) http://www.receiver.vodafone.com/12/articles/pdf/12_01.pdf.

2. John Boyd, "The Only Gadget You'll Ever Need," *New Scientist*, 5 (2005): 28.

3. Roger Silverstone and Eric Hirsch, eds., *Consuming Technologies: Media and Information in Domestic Spaces* (London: Routledge, 1992); Roger Silverstone and Leslie Haddon, "Design and Domestication of Information and Communication Technologies: Technical Change and Everyday Life," in *Communication by Design: The Politics of Information and Communication Technologies*, ed. Roger Silverstone and Richard Mansell (Oxford, UK: Oxford University Press, 1996): 44–74; Daniel Miller, *Material Culture and Mass Consumption* (London: Blackwell, 1987).

4. Mary Douglas and Baron Isherwood, *The World of Goods: Towards an Anthropology of Consumption of Goods* (London: Routledge & Kegan Paul, 1979).

5. Dick Hebdige, *Hiding in the Light: On Images and Things* (London: Routledge, 1988).

6. Miller, *Material Culture and Mass Consumption*.

7. John Agar, *Constant Touch: A Global History of the Mobile Phone* (Cambridge: Icon Books, 2003).

8. Boyd, "The Only Gadget You'll Ever Need."

9. Ilpo Koskinen, "Managing Banality in Mobile Multimedia," in Raul Pertierra, ed., *The Social Construction and Usage of Communication Technologies: European and Asian Experiences* (Manila: University of the Philippines Press, 2007), 48–60.

10. Barbara Scifo, "The Domestication of the Camera Phone and MMS Communications: The Experience of Young Italians," in Kristóf Nyfrí, ed., *A Sense of Place: The Global and the Local in Mobile Communication* (Vienna: Passagen Verlag, 2005), 363–73; Mizuko Ito and Daisuke Okabe, "Camera Phones Changing the Definition of Picture-Worthy," *Japan Media Review* (2003), http://www.ojr.org/japan/wireless/1062208524.php; Mizuko Ito and Daisuke Okabe, "Intimate Visual Co-Presence," presented at *UbiComp 2005*, 11–14 September, Takanawa Prince Hotel, Tokyo, Japan, http://www.itofisher.com/mito/.

11. Ito and Okabe, "Intimate Visual Co-Presence."

12. Chris Chesher, "Neither Gaze nor Glance, but Glaze: Relating to Console Game Screens," *SCAN: Journal of Media Arts Culture*, 1(1) (2004), http://scan.net.au/journal/.

13. Ito and Okabe, "Intimate Visual Co-Presence."

14. Frans Mäyrä, "The City Shaman Dances with Virtual Wolves—Researching Pervasive Mobile Gaming," *Receiver*, 12 (2003), www.receiver.vodafone.com.

15. http://www.pacmanhattan.com/.

16. Nicolas Bourriaud, *Relational Aesthetics*, trans. Simon Pleasance and Fronza Woods (Dijon, France: Les Presses du Reel, 2002).

17. Michael Arnold, "On the Phenomenology of Technology; the "Janusfaces" of Mobile Phones," *Information and Organization*, 13 (2003): 231–56.

18. Judy Wajcman et. al, "Intimate Connections: The Impact of the Mobile Phone on Work Life Boundaries," see this volume. Also see Melissa Gregg, "Work Where You Want: The Labour Politics of the Mobile Office," presented at *Mobile Media Conference* (University of Sydney, July 2007).

19. Jay Bolter and Richard Grusin, Remediation: Understanding New Media (Cambridge, MA: MIT Press, 1999).

20. Margaret Morse, Virtualities: Television, Media Art, and Cyberculture (Bloomington: Indiana University Press, 1998).

21. Larissa Hjorth, "Locating Mobility: Practices of Co-presence and the Persistence of the Postal Metaphor in SMS/MMS Mobile Phone Customization in Melbourne," *Fibreculture Journal*, 6 (2005), http://journal.fibreculture.org/issue6/issue6_hjorth.html.

22. Alex Taylor and Richard Harper, "Age-Old Practices in the "New World": A Study of Gift-Giving between Teenage Mobile Phone Users," in *Changing Our World, Changing Ourselves* (proceedings of the SIGCHI Conference on Human Factors in Computing Systems, Minneapolis, 2002): 439–46; Alex Taylor and Richard Harper, "The Gift of Gab? A Design Oriented Sociology of Young People's Use of Mobiles," *Journal of Computer Supported Cooperative Work*, 12 (2003): 267–96.

23. Douglas and Isherwood, *The World of Goods*.

24. Erkki Huhtamo, "From Kaleidoscomaniac to Cybernerd: Notes Toward an Archaeology of the Media," *Leonardo*, 30(3) (1997).

25. Erkki Huhtamo, "From Kaleidoscomaniac to Cybernerd," 222; cited in Jussi Parikka and Jaakko Suominen, "Victorian Snakes? Towards a Cultural History of Mobile Games and the Experience of Movement," *Games Studies: The International Journal of Computer Game Research*, 6(1) (2006), December, http://gamestudies.org/0601.

26. Parikka and Suominen, "Victorian Snakes? Towards a Cultural History of Mobile Games and the Experience of Movement."

27. Parikka and Suominen, "Victorian Snakes?"

28. Parikka and Suominen, "Victorian Snakes?"

29. Marshall McLuhan, *Understanding Media* (New York: Mentor, 1964).

30. Haddon and Silverstone, "Design and Domestication of Information and Communication Technologies"; Miller, *Material Culture and Mass Consumption*.

31. Daniel Miller and Heather Horst, Cell Phone (Oxford and New York: Berg, 2006), 7.

32. Rich Ling, *The Mobile Connection* (San Francisco: Morgan Kaufmann Publishers, 2004).

33. Leslie Haddon, *Empirical Research on the Domestic Phone: A Literature Review* (Brighton, UK: University of Sussex Press, 1997).

34. David Morley, "What's 'Home' Got to Do with It?" *European Journal of Cultural Studies*, 6(4) (2003): 435–58.

35. Leslie Haddon, "Domestication and Mobile Telephony," in *Machines That Become Us: The Social Context of Personal Communication Technology*, ed. James E. Katz (New Brunswick, NJ: Transaction Publishers, 2003), 43–56.

36. Ling, *The Mobile Connection*, 28.

37. Ling, *The Mobile Connection*, 27.

38. Timo Kopomaa, "The City in Your Pocket," in *Birth of the Mobile Information Society* (Helsinki: Gaudemus, 2000).

39. Huhtamo, "From Kaleidoscomaniac to Cybernerd."

40. Lev Manovich, "The Paradoxes of Digital Photography," in *The Photography Reader*, ed. Liz Wells (London, Routledge, 2003), 240–49.

41. Lisa Gye, "Picture This," paper presented at *Vital Signs* conference (September 2005, ACMI, Melbourne).

42. Cited in Mizuko Ito, Daisuke Okabe, and Misa Matsuda, eds., *Personal, Portable, Pedestrian: Mobile Phones in Japanese Life* (Cambridge, MA: MIT Press, 2005).

43. Arnold, "On the Phenomenology of Technology."

44. Ingrid Richardson, "Pocket Technoscapes: The Bodily Incorporation of Mobile Media," in *Continuum: Journal of Media and Cultural Studies*, 21(2) (2007): 205.

45. Richardson, "Pocket Technoscapes," 208.

46. Robert Hughes, *The Shock of the New* (London: Thames and Hudson, 1981).

47. Matsuda, cited in Ito, *Personal, Portable, Pedestrian*.

48. Richardson, "Pocket Technoscapes."

49. Alvin Toffler, *The Third Wave* (William Morrow: New York, 1980).

50. Dan Tapscott, *The Digital Economy: Promise and Peril in the Age of Networked Intelligence* (New York: McGraw-Hill, 1995).

An iPhone in Every Hand

Media Ecology, Communication Structures, and the Global Village

By Tom Valcanis

Media ecologist Neil Postman once remarked that "A medium is a technology within which a culture grows; that is to say, it gives form to a culture's politics, social organization, and habitual ways of thinking." To what extent has the current "new media" (TV, print, and social and Internet media) created a common globalized media environment and culture?

If one thinks of media in their everyday life, patterns emerge that validate the late Neil Postman's hypothesis—we all have heard variations on the following: "Have you got Facebook?"; "all the news sites are saying ..."; and the ubiquitous "have you heard about so-and-so in the blogs?" Superficially, these examples seem like banal excesses of a leisurely culture with an overabundance of free time to spend on entertainment. However, probing further, it underpins a certain truth that Postman and his colleagues in the Media Ecology Association and scholars frequently cite—that new media technologies do not just *add* to a culture, they *transform* it completely. In doing so, the old ways can only be comprehended in what Marshall McLuhan called "the rearview mirror." Throughout the history of our species, humans have sought to "conquer time and space through speech, art and architecture, through writing and printing, and through various forms of transportation."[1] Through humanity's advancement through technology, we have also made vast changes that have had global repercussions.

In the nineteenth century, the Western world, at the very least, gained access to instantaneous communication technology: the telegraph, the first ever electronic (electrically powered) method of telecommunications. This evolved and expanded with the invention of the telephone, fax machines, radio, television, and innovations such as copper and fiber-optic cable, and satellite communication—all part of the pre-computer mediated communication (CMC) revolution carried over into the "new" media culture that forms an integral part of our modern experience.[2] In the late 1980s, personal computing became more affordable and with it, telecommunications were integrated with this new technology. From this watershed, French philosopher and sociologist Jacques Ellul proposed that the convergence of media and communication technologies (print, video, audio, and telegraph) on the computer has

> ... set up networks in society that have nothing whatever to do with ancient networks or traditional structures ... We cannot continue as before. Simply because the computer is there, we cannot ignore it. When the railroad and the automobile came on the scene, those who wanted could still travel by horseback. But now there is no choice. A businessman cannot acquire a computer just because he likes progress. The computer brings a whole system with it ... the technical system has become strongly integrated ... offices, means of distribution, personnel must all be adapted to it.[3]

The computer has penetrated the lives of almost all people on the planet, arranging them into an interconnected, "retribalized human community within which sight and sound are global in extent," as media scholar Marshall McLuhan noted, which he termed the "global village." The mediation "space" is often now referred to as "cyberspace."[4]

This global village has thrust mankind into a new "information age" or era in which human communication is "growing so fast as to be in fact immeasurable," and as media ecologist, communications scholar, and Catholic priest Walter J. Ong professed, "making human consciousness something other than what consciousness used to be" instead "moving into ... a situation where, in principle, everything that is known or has been known can be made accessible to everyone everywhere everytime."[5] If this phenomenon is truly global and we take the premises of Postman's axiom as true-to-fact, then there indeed exists a globalized culture, which these new mediums are shaping and re-shaping from day to day and even hour to hour.

This article seeks to explain, using media ecology as an analytical framework, whether globalization and the technological "information age" brought on by new media convergences (Information Communications Technology and the computer, Internet, social media—YouTube, Facebook, Skype, etc.—and smartphones) that we currently experience are a transformative and total cultural phenomenon. Before one can determine the how and why, one must first define, contextualize, and reify precisely what they are looking at. Do they really have a "new" stranglehold on culture or are they just extensions of what we have experienced previously?

A "New" Media Ecology

Media ecology is an exploration into, as Marshall McLuhan defined it, "the matter of how media of communication affect human perception, understanding, feeling, and value; and how our interaction with media facilitates or impedes our chances of survival. The word ecology implies the study of environments: their structure, content, and impact on people."[6] Media ecology serves as an interdisciplinary approach that converges on studies of language, media analysis, education theory, radical constructivism, communication theory, philosophy of mind, anthropology, and even humanistic, non-Aristotelian epistemologies such as (the itself) interdisciplinary General Semantics pioneered by Alfred Korzybski in 1933 with the publication of *Science and Sanity*. Essentially, media ecology can be styled as the academic study of communications and media technology and its impact on human affairs.[7] Professor Lance Strate, Communication and Media Studies at Fordham University, suggests that "we need to study the new ways that we communicate in the present. And, if we want to understand the present, we need to put it into historical context."[8] Media ecology is distinct as a field of scholarship and analysis as much as it is interdisciplinary and reflexive. However, there are many approaches to theorize and explain a "common global media environment and culture."

Though media ecology could be viewed as simple media criticism, it is not. Scholars like Robert McChesney or Noam Chomsky analyze the ownership of the media concerns and whether it affects the production, exhibition, and distribution of content. Media ecology takes a broader view, concentrating on media technologies and their place in shaping society.[9]

It views media as a culture all into itself, influencing the overall (global) culture and actors and viewers in the media as (a) "cultural production," much like French sociologist Pierre Bourdieu's theory of *The Field of Cultural Production*.[10] Though incomplete (and limited mostly to literature and art and, only briefly, journalism), the "field"

> ... theorizes interconnections between different areas of endeavor, and the degree to which they are *autonomous* of each other. The major fields Bourdieu tends to write about are the economic and political fields, and a composite of the two, which he calls 'the field of power'; the educational field; the intellectual field; and various cultural fields, including the literary field, the artistic field, the scientific field and the religious field.[11]

These fields are all bound together by capital—be it creative or monetary—which provides the worth, promulgation, and influence of the content created. Though useful in understanding in where the content of the media is "coming from" and why, it falls flat in explaining media both as a culture and as a culture-producing entity. But what are these new mediums that construct a global culture?

New Media as a "catchcry" or "buzzword" is a higher-level abstraction that attempts to convey computers (from the desktop to the hand-held) connected to the Internet to carry images, audio, video, and text, as well as realtime telecommunications as a medium. These all converge and overlap one another—one text can be connected to another via hyperlinks over the World Wide Web, which could also have video content embedded within that can be produced by virtually anyone and seen by the same amount of people; these texts are no longer constrained by the producer/consumer divide—they are in constant change, are interactive, and are amorphous in nature.[12] As McLuhan and Fiore provide that "all media work us over completely ... [being] so pervasive in their personal, political, economic, aesthetic, psychological, moral, ethical, and social consequences that they leave no part of us untouched, unaffected, unaltered," so too our culture on a global scale falls under this umbrella.[13]

If we take new media as the conduit for a globalized culture "best understood as the meta process of an increasing, multidimensional worldwide connectivity," then the media as a culture shares points of habit and organization around the globe that contribute to the overall structure of this speculative "global village" or "network society."[14] The most obvious and fundamental medium enabling a network society is the Internet, accessible via computer or device which resembles its function.

No Space in Any Time

Media was fundamentally developed to traverse the limitations of a preliterate, oral culture—orality could only sustain communication within a predetermined perceptual space, the immediate surroundings of the speaker and the listener.[15] We have sought, as a species, to reduce the tyranny of communicating information over vast distances first by substituting the "ear for the eye" in writing and print and then increase the speed of transmission of these messages by inventing more and more sophisticated transportation mechanisms to theoretically extend our perceptual space to reaches hitherto unknown and beyond.[16]

These methods of transport were then supplanted by electricity and the telegraph, telephone, and eventually wireless technologies such as radio and television. These mediums enabled one to broadcast information to a vast amount of people instantaneously and across great distances. In the computer age, these forms have melded into a seemingly singular form of telecommunications, information media, commerce, and even cultural production as well as a new type of user-generated (i.e., non-professionally) and read "social media."

In the "old days," to fashion a metaphor, the medium was the slip of paper and the transmission method was horse and cart; the new global medium is the multimedia computer and the transmission method is the Internet.[17]

The computer and the Internet—the dominant form of Information Communications Technology (ICT) on the entire planet—can be likened to a device that can simultaneously write, publish, and be read from; a conduit from which audio and video can be captured, edited, and displayed; and a terminal from which all these elements can be transmitted to other either over existing electronic communication networks or wireless. The metaphor that is used to convey this "space" in which the computer traverses is *cyberspace*—as Rosanne Stone defines:

[Cyberspace] can be characterized by virtual space—an imaginary locus of interaction created by communal agreement. In its most recent form, concepts like distance, inside/outside and even the physical body take on new and frequently disturbing meanings.

Though Strate et al. emphasize that cyberspace is

the conceptual space where words, human relationships, data, wealth and power are manifested by people using CMC technology ... but is not identical with communication through computer media but rather the context in which such communication occurs.[18]

Though thought to emerge in the United States, cyberspace is a metaphorical landscape that encompasses the entire world over and even beyond our atmosphere, as satellites are used to facilitate this "space" and "time." The media is not only everywhere and everytime as Ong posits but also the transformative impetus toward a globalized media culture (or more specifically, media as culture.)

Globalization is "an inevitable element of our lives. We cannot stop it any more than we can stop the waves from crashing on the shore."[19] Globalization grew out of a desire for individuals to open up new markets and methods to handle information and information flows across great distances. Globalization as an abstract concept has created a new symbolic and semantic environment that "reaches right around the globe, which is organized, in very large part, by media transnational corporations."[20] This environment also has its constituent parts and actors—human beings—that arrange themselves not into societies but into "networks," as a structure, the communications between the nodes in the network termed as "flows."[21] The "flows" required massive amounts of ICT infrastructure development globally to prop up these networks and facilitate cyberspace. For example, international telephone connections to and from the United States grew 500% between 1981 and 1991, from 500 million to 2.5 billion.[22] The computer was in its infancy in 1991 as a global networking conduit, but in the latter part of the twentieth and early part of the twenty-first century, the computer as well as the computer-enabled "smartphone" took off as a part of a media culture. In the beginning, email and hyperlinked text-based web-pages were the most obvious form of content to be found and transmitted online. This expanded into the embedding of images, audio, video, and interactive games, leading to the development of user-oriented publishing such as the weblog or "blog."[23]

All throughout this period, non-Web-based media also emerged in the form of instant messaging with applications such as ICQ, Internet Relay Chat clients, and MSN Messenger being freely available for download and use. Transmission of larger files (movies, music, and books) was mediated by peer-to-peer networks such as Napster,

Kazaa, and BitTorrent. In 2005, "Web 2.0" arrived. This "new" web was driven by user-generated content and social media, distinct from the old "static" Web 1.0 to "dynamic" web-pages that could be altered by end-user input—much like the collaborative and user-alterable "wikis," made famous by Wikipedia, that can store easily accessible and hyperlinked information with embedded video, audio, and images. These new forms of "social media" are open to all that can connect to them; they have a dual function of producing/consuming user-made content and for CMC-based social networking; and unlike traditional media, it has no physical space and may be read in fragments and/or nonsequentially.[24]

With the promulgation of wireless technologies such as Wi-Fi or 3G WiMax (high-speed wireless voice and data transmission) being embedded into "smartphones"—i.e., mobile phones that act as "miniature, mobile" computers with abbreviated (or in some cases, full-featured) applications. These so called 'apps' mirror similar programs and the communications ability of desktop or laptop computers—the most popular being the Apple iPhone along with Research in Motion's BlackBerry series, Nokia, and Google's operating system Android, which powers Samsung, LG, and other phones. Mobile computing via smartphones are a new form of media that deals directly with the "moving human body and the ecological interrelationships among the virtual space of the Internet, the enclosed space of the installation, and the open space of everyday life."[25] As such, mobile smartphones now feature GPS technology to enable other users to locate their whereabouts through websites and other social media platforms.

So what measurable impact have these new media technologies as cultural devices actually had on globalization in forming a new global media culture?

It has, for the most part, transformed the global culture, at its fundamental essence, into a *participatory culture* that sees the computer not as a new "steam engine" but rather something much more revolutionary in terms of human organization on a global scale—the new "mechanical clock."

It Is Everyone's Turn, All the Time

If the technology is the medium in which a culture grows, the interactive and user-oriented nature of these technologies have given rise to a participatory and "mash-up" culture in which the ways of producing and accessing content are deconstructed, uploaded, mixed, converged, and reconstructed through computers and smartphones mediated by online platforms; it becomes a "participatory culture" as defined by media scholar Harry Jenkins:

> [That] contrasts with older notions of passive media spectatorship. Rather than talking about media producers and consumers as occupying separate roles, we might now see them as participants who interact with each other according to a new set of rules that none of us fully understands. Not all participants are created equal. Corporations—and even individuals within

corporate media—still exert greater power than any individual consumer or even the aggregate of consumers ... Consumption has become a collective process.[26]

This participatory culture is explained as being part of a continuum of "people moving through time, [with] each group or generation of people possessing a distinct sense of self" which superficially can be determined by the explosion of users of social media platforms such as video site YouTube, Facebook, or "short form Hogging" site Twitter; Facebook itself has over 500 million users, which is approximately 1/12 of the entire population of the world.[27] Even though this does not explain whether this participatory culture has given rise to new methods of globalized culture or habitual ways of thinking, rather it could be seen as an extension of the Habermasian "public sphere" that has "re-tribalized" itself into smaller subsets of societal or subcultural networks instead of a "traditional" citizenry gaining access to democratic institutions via the media "fourth estate."[28] The computer at the very least and the new "participatory culture" at the very most can be likened to the revolutionary power of the mechanical clock.

A computer is built on a time-telling function—time regulates the processing of information by creating a sense of "dramatic, fictional or symbolic time as well as a sense of past, present and future."[29] Computers, like clocks, are self-operating machines; they manufacture no physical products. They are geared toward production rather than distribution; community over the product; service over commodity and creating "economically effective links between people and information" such as regulating starting and ending times for social/economic/political engagements; and enforcing deadlines for the furnishing of media owners for cultural production (such as copy, films, images, or other marketable material).[30] For example, all people across the known world began to

"... work, sleep and eat by the clock" and began to "regulate their actions by this arbitrary measurer of time, the clock was transformed from an expression of civic pride into a necessity of urban life ... the computer too has changed from a luxury to a necessity for modern business and government.[31]

Digital, computer-measured time is not just a quantitative measurement but a concept—represented as a sequence of numbers—digital *information*. This conceptual "cyberspace" also gives rise to "cybertime" signaling the end of space-conquering societies in lieu of "time-conquering" information societies. "Cybertime" is a polychronic time, which involves many things occurring simultaneously—McLuhan argues that such a time is "characteristic of many non-Western cultures and increasingly, of electronic cultures."[32]

This lends itself to the concept of the space nonspecific "cyberweek" in which timezones are made irrelevant by computer-mediated communication and the time-as-information society.[33]

Leading from that, sociologist Manuel Castells' theory of the information society stems almost directly from the computer-as-time binder metaphor where, in the 1980s, "the information revolution began with a restructured capitalism ... creating a global society that is connected by networks."[34] Globalization, according to sociologist Anthony Giddens, is "the intensification of worldwide social relations which link distant localities in such a way that local happenings are shaped by events occurring many miles away and vice versa" that is itself mediated by the mechanical clock, universal Gregorian calendar, and, of course, ICT and CMC technology.[35] For example, the microblog Twitter opted to change its local time for maintenance in the United States to allow Iranian anti-government protesters to post or "tweet" their stories of abuse and army crackdowns following the 2009 disputed election as foreign journalists were barred from entry to the country.[36] One such ubiquitous application of this new media technology is the social networking site Facebook.

Facebook's primary purpose is to "share information with people you know, see what's going on with your friends, and look up people around you."[37] On the site itself, we see the ability to write short- or long-form blogs, upload pictures, videos, links to other websites, and integrate other social media and "like" cultural products into the "timeline" of social interaction, which is the central focus of the site. If we take the concept of cybertime and Bordieu's position that "a social environment consists of a multiplicity of social fields in which agents produce practices" of which Facebook is a "social field," its "agents" the users.[38] There is a distinct extension of the "social mind" insofar that Facebook creates a "present of past and future things"—one can reminisce with friends in their network about past events and organize future events by inviting networked friends.[39] For example, a conversation can happen in real time using the instant messaging system or in a bulletin board system format of comments on posted material forming "threads"—"a lively back and forth of discussions that could have lasted days, weeks years ... scrolling down the screen [gives the illusion] as if they were taking place in real time, which for the reader watching them flow past on the screen, they are."[40] Naturally, Facebook is a computer mediated experience that requires the use of one or an Internet-enabled smartphone. It is popular and enables us to maintain relationships, pass time, become part of a community, and entertain us, much like traditional media did in the past, for example, spending time with friends to see a movie.[41]

Facebook is only part of an autonomous, automatic, and self-augmenting network system fuelled by the interaction of those in a network society, according to Ellul.[42] But this brings us to a larger ecology of media that encompasses our use of media as the message and media as metaphor—media as language itself.

The (In)Conclusion

Postman in his magnum opus *"Amusing Ourselves to Death"* wrote the easiest way to see through a culture is to "attend to its tools for conversation."[43] Currently, all our conversation, save for face-to-face contact, is mediated, at some level, by computers and the Internet—the tools—and the conversation—the exchange of messages—is happening globally in which any user of a computer is theoretically part of this "globalized conversation."[44] But what is the nature of the language of this conversation—the "driver" of conversation that makes it possible?

The Sapir-Whorf hypothesis presents the formation of language is "not merely a reproducing instrument for voicing ideas, but rather itself is a shaper of ideas."[45] The computer and the Internet and all its various convergent and multimedia forms not only have produced new platforms for communication, they have, in fact, shaped a new way of organizing and regulating ideas: the way humans interact with one another and conduct their business, their politics and their education of future generations.

For example, politicians embrace social media not to appear "with it" and appeal to a younger audience (or at least give that impression) but rather as a political necessity as the media as culture shifts toward a "global village" based on CMC, ICT, and social media.[46] As soon as television and computers were made available for classrooms, teachers began to include them in their curriculum as a learning device as well as their proper use.[47]

It is questionable to conclude if this culture really has been radically "transformed." However, if the medium is the message and these mediums change over time, the cultural changes are also tangible and material. For one to connect to the Internet, one must purchase a computer or smartphone and an Internet connection (be it wireless or cable)—a computer or like device is now a near universal fixture in Western homes, much like the television and telephone before it.[48] This global communication culture has undoubtedly had a material impact on our politics, our economics, and our cultural production and reception. It has "given [to us] as it has taken away" insofar that we "worship" technology as Postman says, but it is almost undeniable; we as humans are now completely different as a people, as a society, and as a networked global "village."[49]

Notes

1. Innis, H.A. *The Bias of Communication,* University of Toronto Press: Canada, 1951. p. 161.

2. Strate, L., Jacobson, R.L., and Gibson, S.B. "Surveying the electronic landscape: an introduction to communication and cyberspace" in *Communication and Cyberspace: social interaction in an electronic environment* (also ed.), Hampton Press: Cresskill, NJ, 2003. p. 5.

3. Ellul, J. *The Technological Bluff,* Eerdmans Publishing Co.: Grand Rapids, MI, 1990. p. 9.

4. Dixon, K. *The Global Village Revisited: Art, politics and television,* Lexington Books: Plymouth, UK, 2009. pp. 8–9.

5. Ong, W.J. "Information and/or Communication" in Farrell, T. J. and Soukup, P. A. (eds) *An Ong Reader: challenges for further enquiry,* Hampton Press: Cresskill, NJ, 2002. p. 517.

6. McLuhan, M. *Understanding Me: Lectures and Interviews* (ed. McLuhan, S. and Staines, D.) MIT Press: MS, USA, 2004. p. 271.

7. Gencarelli, T.F. "Neil Postman and the Rise of Media Ecology" in *Perspectives on Culture, Technology and Communication* (ed. Lum, C. M. K.) Hampton Press: USA, 2006. p. 203.

8. Strate, L. "The future of consciousness" in *ETC: A Review of General Semantics* (Jan 1, 2009). [online]

9. McChesney, R.W. "Media Concentration" in Bucy, E.P. (ed.) *Living in the Information Age: A New Media Reader,* Wadsworth: Southbank, AU, 2005. p. 111.

10. Hesmondhalgh, D. "Bourdieu, the media and cultural production" in *Media, Culture and Society,* (Vol. 28, 2006) p. 211.

11. Hesmondhalgh, "Bordieu", p. 212.

12. O'Sullivan, J., and Heinonen, A. "Old Values, New Media" in *Journalism Practice* (Vol. 2, No. 3, 2008) p. 359.

13. McLuhan, M., and Fiore, Q. *The Medium is the Massage,* Penguin: UK, 1967. p. 26.

14. Hepp, A. "Translocal Media Cultures: Networks of the Media and Globalization" in *Conference Papers—International Communication Association,* (2006 Annual Meeting), p. 7.

15. Strate, L. 'Cybertime' in Strate, L., *Communication and Cyberspace: social interaction in an electronic environment* (also ed.), Hampton Press: Cresskill, NJ, 2003. p. 363.

16. McLuhan, M. *Understanding Media: the extensions of man,* Routledge: UK, 1964. p. 99.

17. Strate, et al. 'Surveying the electronic landscape', 2003. p. 5.

18. Ibid, p. 4.

19. Fairclough, N. "Language and Globalization" in *Semiotica* (Vol. 1, No.4, 2009) p. 325.

20. Webster, F. *Theories of the Information Society,* Routledge: UK, 2002. p. 72.

21. Hepp, A. "Translocal Media Cultures", 2006. p. 3.

22. Webster, *Theories,* 2002. p. 73.

23. Goggin, G. "The Internet, Online and Mobile Communications and Culture" in Cunningham, S. and Turner, G. (eds.) *The Media and Communications in Australia* (3rd ed.) Allen and Unwin: NSW, Australia, 2010. p. 245.

24. Serazio, M. "(New) Media Ecology and Generation Mash-Up Identity: The Technological Bias of Millennial Youth Culture" in *Conference Papers—National Communication Association* (2008) p. 7.

25. Pedersen, I. "No Apple iPhone? You Must Be Canadian—Mobile Technologies, Participatory Culture, and Rhetorical Transformation" in *Canadian Journal of Communication* (Vol. 33, No. 3, 2008) pp. 493–494.

26. Pedersen, "No Apple iPhone?", 2008. p. 494.

27. Zuckerberg, M. "500 Million Stories", *The Facebook Blog* posted July 2010. [online]

28. Habermas, J., Lennox, S., and Lennox, F. "The Public Sphere: An Encyclopedia Article" in *New German Critique,* No. 3 (Autumn, 1974) p. 54.

29. Strate, L. "Cybertime" in *Communication and Cyberspace: social interaction in an electronic environment* (also ed.), Hampton Press: Cresskill, NJ, 2003. p. 363.

30. Rushkoff, D. "The Information Arms Race" in *Communication and Cyberspace: social interaction in an electronic environment* (ed. Strate, L. et al.), Hampton Press: Cresskill, NJ, 2003. p. 355.

31. Strate, "Cybertime", 2003. p. 364.

32. Ibid, p. 369.

33. Laguerre, M.S. "Virtual Time" in *Information, Communication & Society* (Vol. 7, No. 2., Jun. 2004) p. 229.

34. Allan, K. *Contemporary social and sociological theory: visualizing social worlds,* SAGE Publications: UK, 2010. p. 176.

35. Allan, *Contemporary social theory*, 2010. p. 265.

36. Grossman, L. "Iran Protests: Twitter, the Medium of the Movement", *TIME Magazine* 17 June 2009. [online]

37. Sheldon, P. "Student Favorite: Facebook and Motives for its Use" in *Southwest Mass Communication Journal* (Spring 2008) p. 40.

38. Rawolle, S., and Lingard, B. "The mediatization of the knowledge based economy: An Australian field based account" in *Communications* (Vol. 35, 2010) p. 269.

39. Strate, "Cybertime", 2003. p. 340.

40. Ibid.

41. Sheldon, "Student Favorite", 2008. pp. 44–47.

42. Strate, L. *Echoes and Reflections: On Media Ecology as a field of study*, Hampton Press: Creskill, NJ, 2006. p. 73.

43. Postman, N. *Amusing Ourselves to Death,* Methuen: London, UK, 1987, p. 9.

44. Strate, L. *Echoes and Reflections*, p. 51.

45. Bois, S.J. *The Art of Awareness: A textbook on General Semantics and Epistemics*, W.C. Brown Co.: USA, 1973. p. 157.

46. Gurevitch, M., Coleman, S., and Blumler, J.G. "Political Communication—Old and New Media Relationships" *The Annals of the American Academy of Political and Social Science* (Vol. 625, 2009) p. 164.

47. Postman, N., and Weingartner, C. *Teaching as a Subversive Activity,* Penguin Books: Middlesex, UK, 1969. p. 28.

48. Strate, *Echoes,* 2006. p. 100.

49. Postman, N. *Technopoly: the surrender of culture to technology,* Random House: New York, USA, 1992. p. 29.

Mobile Telephony and Mediated Ritual Interaction

By Rich Ling

Rituals are not only enacted locally; they are also mediated. Cohen et al. (2007) describe a bris (circumcision ceremony) that took place in Israel. A bris takes place on the eighth day after a male child is born. In this case, the boy's father, serving reserve duty in the Israeli army, was unable to attend in person. However, he was able to participate via a mobile video connection. The people on one end of the transmission were participating in a traditional bris. On the other end, the father was able to remotely play his part. Indeed, the father had an active role. He was not simply a viewer. Thus, there was a ritual taking place in two locations, mediated by the mobile video link. Such a situation is not new. Standage (1998, pp. 128–129) discusses marriage ceremonies that were carried out by telegraph in the 1800s. Although such situations are not common, they indicate that distance does not necessarily get in the way of a good ritual.

In the previous chapter I focused on how mobile communication influences and often disrupts co-present situations. In this chapter, I will focus more directly on the way that ritual is played out via the actual mobile communication events. In this respect, I am taking a step away from Collins's assertion that co-presence is an essential element in ritual. The main theme here is that groups are able to engender cohesion with the aid of mobile communication. The mobile phone can be used to destroy cohesion or to build it. However, as we will see in the following chapter, the balance seems to be tipping in the direction of mobile communication's supporting the development of cohesion in small groups.

Mediated Ritual

For Durkheim, as I noted above, ritual was, per definition, a co-present activity. Durkheim worked in an era where interpersonal interaction was only rarely mediated. In addition, the objects of his study—Australian Aborigines—may never have come into contact with any form of mediation technology. Goffman worked in an era and in locations with greater access to telephony, but he was for the most part interested in co-present "face engagements" while still recognizing that mediated interaction is an arena in which interaction can take place (1963a, p. 89, n. 12). Collins (2004a, p. 23) asserts that ritual interaction can be developed only in co-present situations, and (with some justification) that it is not possible to develop the sense of mood and effervescence needed to carry off a successful ritual via a remote hookup. Collins discusses, for example, the hypothetical idea of a wedding conducted over a remote link. He notes that the inability of the participants to provide direct feedback would hinder the dynamics of the event. Indeed, events such as weddings (or the bris described above) are rooted in co-present tradition, and it is in that realm that they have their existence (Rice 1987).

It is clear that co-present rituals are powerful motors for the development and maintenance of social cohesion. To quibble with this is pointless. In the broad sweep of social interaction, it is as Collins suggests. Weddings, sporting events, funerals, parties, meetings, concerts, and other events we attend in person are the situations in which we are able to best forge social links. With marginal exceptions, all other forms of mediated interaction pale in comparison to the power of co-present interaction. That said, mediated interaction is also a form of contact through which social bonds can be nurtured. Mediated interaction can influence the staging of broader co-present events. For example, it can be used by operatives arranging a political rally to check on details, orchestrate the timing of the event, and fine-tune the use of effects during the actual rally.

Participants can use mediated interaction to anticipate the event and also to recap it afterwards (Ito 2005b). Mobile communication can also be used by participants during a concert, a rally, or a service. For example, participants can broadcast the excitement of being at a concert to others who are not there by means of a "cellcert" (Watkins 2005). Mediated interaction can take its point of departure from the symbols and routines used in the co-present ritual. These can be developed, reinterpreted, or revitalized in the subsequent mediated interaction. The catch phrase from the theater piece or the sermon can be restated and reinterpreted in the phone calls afterward, or the big play in the football game can be dissected in the text messages exchanged between the fans.

Apart from large-scale "Durkheimian" rituals, mediated communication can influence the more Goffmanian forms of interpersonal interaction. As will be described below, the device can be used to facilitate and augment smaller-scale interpersonal interactions. Finally, various forms of ritual can be developed entirely within the realm of mediated interaction. Ways of greeting, forms of argot, and even phrasing can be characteristic of a mediation form. In addition to the communication of information, this meta-language can serve to underscore that the communication is taking place via a particular kind of mediation and that in itself is a sign of inclusion.

While co-presence is a major venue for the development of social cohesion, it is also possible to assert that some of the same can be achieved via mediated forms of interaction. Mediated interaction can enhance the broader co-present forms of interaction and can also function in its own right as a means through which members of a group can engage one another and develop a common sense of identity. Indeed, as will be discussed in the next chapter, the directness and ubiquity of the channel can lead to the tightening of social bonds within a group.

These notions will be examined using several different examples of mediated interaction via mobile phone, including greeting sequences, romantic interaction, texting, jokes, repartee, and gossip.

Forms of Mediated Ritual Interaction

Pre-Configured Greetings

Greetings may be the most thoroughly examined of the various mediated ritual forms. Goffman examined them (1967), as did Sacks, Shegloff, and Jefferson (Schegloff and Sacks 1973; Sacks et al. 1974; Schegloff et al. 1977). Goffman suggests that greetings are a way to assure the participants that the "relationship is still what it was at the termination of the previous coparticipation" and that greetings "serve to clarify and fix the roles that the participants will take during the occasion of talk and the commit participants to these roles" (1967, p. 41, n. 30).

Telephonic greetings—for example, "hello" (Martin 1991, pp. 155–163; Bakke 1996; Marvin 1988; Fischer 1992, pp. 70–71)—may be the most ritualized portion of the common telephonic situation. But there were difficulties lurking behind what we now see as a simple greeting. In early telephone conversations in the United Kingdom, there was a need to work out the status differences of the speakers that previously had been handled visually (Fischer 1992, p. 70). It was seen as an affront to an upper-class person to be addressed by a lower-class person with a familiar "hello." Thus, elaborate forms of address were developed, such as "Mr. Wood of Curtis and Sons wishes to talk with Mr. White" or "Hello, this is the Jones residence, Samuel speaking." Time has not been kind to such overblown expressions. By the time Sacks, Schegloff, and Jefferson (1973, 1974) examined greetings, the following was typical: "Hello." "Hi, this is John." "Hi John, how are you?"

Laurier (2001) has discussed how greetings have changed to include geographic location. Beyond geographical information, the mobile telephone has also changed greeting rituals by doing away with the need for the interlocutors to identify themselves.

Observation A man was standing in a store on a Saturday morning. He was examining some items in the store when his phone rang. He pulled the phone out of his right jacket pocket and looked at it long enough to activate it and read the caller ID in order to know who was calling. Then he raised the phone to his ear and, instead of the traditional greeting sequence, said "Are you ready?" (pause) "OK." There followed a short conversation with two or three turns. The conversation was quickly over and the man replaced the phone in his pocket.

The lack of ceremony in the opening bespoke a call between two persons who were in frequent contact. This is also a good example of pre-configuring the greeting sequence. There was no need for introductions, formal interactions, or the so-called double introduction (Schegloff and Sacks 1973; Sacks et al. 1974). The mobile phone reduces or even eliminates the need for any form of opening, since the person making the call is calling to a specific individual, and the person receiving the call usually has the more commonly used numbers in his or her name register, meaning that the name of the caller appears via caller ID. Since both partners are clued into this, it is perhaps seen as artificial to go through the sequence of "Hello, this is Rich calling; may I please speak to John?" This is particularly true if the interlocutors are in a connected state. It is, for example, possible to speculate that the man in the observation shown above was only doing a short errand while waiting for his wife to finish another task. The call was meant simply to alert the man to his wife's current status with respect to the broader project of shopping on Saturday. They were so engrained in each other's situation and so aware of the other's general context that the abbreviated messages were easy to unpack (Miyata 2006).

If the telephonic interlocutors are not intimate, a more extended greeting might be called for—one that allows the partners to rearrange their mental furniture in order to place themselves into the context of the call: "Oh, is that you, Sally? It is good to hear from you. Is Dick doing OK? What about Spot and Puff?" If this were the case, then the two interlocutors would have to go through a process of establishing the context of the call and deal with the issues outlined by Goffman. They would need to remind themselves as to the nature of the interaction with the somewhat more distant person. In the case of the call reported in the observation, there was no need for that, since the call was a kind of "connected presence" (Licoppe 2004). Speaking of textual interaction, Ito and Okabe (2005) describe how a mobile phone can be used to give a "virtual tap on the shoulder":

> These messages define a social setting that is substantially different from direct interpersonal interaction characteristic of a voice call, text chat, or face-to-face one-on-one interaction. These messages are predicated on the sense of *ambient accessibility,* a shared virtual space that is generally available between a few friends or with a loved one. They do not require a deliberate opening of a channel of communication but are based on the expectation that one is in "earshot." … As a technosocial system … people experience a sense of persistent social space constituted through the periodic exchange of text messages. These messages also define a space of peripheral background awareness that is midway between direct interaction and noninteraction. (Ito and Okabe 2005, p. 264; emphasis added)

Although Ito and Okabe are describing interaction via text, many of the same points apply to the man described in the observation. If I interpret the situation correctly, there was a kind of ambient accessibility between the two, and there was a general background awareness that the call would come at some point within a particular time frame.

It is also instructive to consider the loose greeting in the observation. As a child, in the age previous to mobile telephony, the man probably was instructed on how to answer the phone using a variation of the more formal introduction where greetings and names are exchanged. However, since the advent of caller ID and the direct interpersonal calling implied by mobile telephony, the more decorative aspects of the greeting have obviously been jettisoned. Had the man used a full formal greeting sequence (including phrases such as "This is John Smith" and "To whom do I have the pleasure of speaking?"), it would been seen as strange. Just as we perhaps greet work colleagues once in the morning, but then do not need to repeat the greeting through the day as we meet them in the hallway or in the lunch room, the man in the store described here obviously did not feel the need to re-greet his interlocutor. Thus, while the initiation of the phone call may have seemed abrupt to other's ears, to behave otherwise may not have been appropriate in the context of that dyad.

The greeting sequence in the observation reported above was certainly a more streamlined version of the traditional greeting sequence. But if I read the situation correctly, the greeting that was proffered, however perfunctorily, may have been the most correct one. It gave evidence to the person calling that the man who took the phone was paying attention and providing the appropriate amount of focus. The man in the store gauged the mood of the interaction—slack time while out shopping on a Saturday—and paid it the proper heed. In addition, the interaction allowed for the further interaction between the two persons.

In some respects, the interaction in the observation is boring. Not much happened. A middle-aged man received a call on his mobile phone, grunted a couple of utterances before hanging up, and continued on his way. But in fact some important things happened. The observation seems to indicate that he maintained the flow in an ongoing interaction with a well-calibrated style, using the appropriate symbolic devices. Had he been much more or much less enthusiastic, he would have sabotaged the goal of the telephone call. As it was, he managed to account for himself and his actions. Was this a ritual interaction? If we tone down the definition of a ritual interaction, then we can assert that there was mutual engagement in the continuing interaction between the man and his interlocutor. Thus, though it was not done with the accompaniment of drums or with intense Durkheimian effervescence, the interaction fulfilled Goffman's notion of mundane yet ritualized activities.

Using a slightly different test as to whether this was a ritual, we can speculate as to whether the interaction could have failed. Again the answer is Yes. The man could have ignored the call; he could have taken the call but not said anything; he could have flown into a rage. In each case, the interaction would have failed as a ritual. As it was, however, he carried it off.

Negotiation of Romantic Involvements

Mobile communication is used in the cause of romance to support co-present interaction. It allows people to exchange comments and endearments when they are not physically together.

It is clear that the mobile phone has affected romantic involvements. In an oddly unbalanced finding, material gathered in Norway shows that 50 percent of teen girls and 32 percent of teen boys flirted by mobile phone at least once a week.

Within a closed group, romance is one of the more ritualized forms of interaction. Indeed, the negotiation of a romantic interest is perhaps the quintessential small-scale ritual. When two people are entering into a relationship, there are hints, glances, and staged interaction. Nearly every word or gesture is interpreted. Various techniques and devices are drawn upon in this small-scale Goffmanian ritual—for example, wearing special perfume, smiling and laughing at comments, the offering of a rose, the planning an intimate meal. If nuanced interaction does not carry the day, a broad hint or a risque double entendre may. After the courting has begun in earnest, there are various expectations placed on the partners as to their mutual availability, the forms of attentions, and the kind of interaction that is to be expected.

Collins (2004a, p. 230) discusses flirting as an interaction ritual chain. He discusses the initiation of flirting via, for example, touching feet underneath the dinner table (ibid., p. 240). Goffman (1971) discusses how the "tie sign" of holding hands helps to define the couple for each other and in the public eye. It is beyond question that co-presence is, in the vast majority of cases, an essential element in the development of a romantic relationship.

As Collins notes, the assembly of the group (in this case often a dyad), the mutual focus of attention, the establishment of a mood, and the barrier to outsiders are all elements that are necessary for the establishment of coupled solidarity. While physical co-presence is often a key element at various points in the development of a romantic relationship, mediated interaction has also been a discreet aspect of what he calls "linking" in the courting process (Collins 2004a, p. 193). Messages and "back-channel" communication have often played a part; indeed these messages are often the handmaiden of love. It is in this way that important information on the other's availability and one's own intentions are made explicit. In literary history, Romeo's ill-timed notes to Juliet and the various missives between Darcy and Elizabeth in Jane Austen's *Pride and Prejudice* illustrate this form. In my own past, passing notes in class (folded in a particular origami-like way) was how nascent couples started a relationship. Thus, there has long been the use of mediation technology in these enterprises. The telegraph (Standage 1998, pp. 134–136), the traditional telephone (Fischer 1992, p. 234; Marvin 1988, pp. 67–68), and now the mobile phone (Andersen 2006; Byrne and Findlay 2004; Ellwood-Clayton 2003; Fortunati and Manganelli 2002; Habuchi 2005; Kim 2002; Ling 2004b; Praitz 2006; Solis 2007) have been used in the exchange negotiation of "love projects" (Praitz 2006).

The initial stages of a relationship are always troubling, particularly for teens. Determining the other's suitability and level of interest and the kind, amount, and tempo of self-disclosure that is appropriate are all problematic. The way with which to deal with these issues has changed. The cycles of being in the public square at the appropriate time and promenading in the accepted places was replaced by interaction by telephone. This again has changed with the adoption of the mobile phone. Material from group interviews in Norway reveals that after meeting a possible love interest, the hopeful couple exchange mobile phone numbers and start a more or less extended and more or less explicit series of text messages (Ling and Helmersen 2000; Ling 2000; Andersen 2006). Rather than synchronous interaction by telephone, in which it is perhaps easy to make a false step, the couple engage in the exchange of carefully edited text messages. As they gain their footing, the frequency of the messages might increase. This indirect form of interaction allows the individuals to cover over some of their more obvious character flaws and to move through the preliminary stages of a relationship in a more deliberate way. It also is a way of saving face in the case of giving or receiving rejection.

> **RITA,** 18: … if you meet a guy when you are out for example, then it is a lot easier to send a message instead of talking like. Somebody you don't really know. It is more relaxed.
> **Anne,** 15: It is easier to tell if you like a person.
> **Interviewer:** Via SMS?
> **IDA,** 18: Then your voice will not either shout or disappear. You need time to think [when constructing your messages].

SMS also makes it possible to contact the potential partner without the fear of being directly rejected, something that is more difficult to manage in voice-based interaction. Andersen (2006, p. 45) reports this exchange:

> **ESPEN:** It is super dumb if you call and you hear that the hag is really uninterested, that she just doesn't say anything.
> **CATO:** Embarrassing when it is completely quiet.
> **ESPEN:** "Why are you calling me?"—you know, that is not cool.

Further, it is important that the message or messages used when flirting have a kind of deniability. This is afforded by playing on humor. Teens note that they send messages that are on the "fresh" side, such as this one reported by Andersen (ibid., pp. 53):

> Im an alian and I've transformed myself in to your phone. While you're reading this Im having sex with your finger. I know you like it because you're smiling;)

If the appeal does not find approval, the author can simply say that he was just kidding and avoid loss of face.

The timing of the message receipt/response cycle is also important. In parallel with how many times a phone is allowed to ring, it is possible to respond to a text message too quickly or too slowly. If the game is played successfully, the two individuals are drawn into the same humor with regards the potential for the relationship. As this happens, the frequency of messages might increase and they might move into more synchronous forms of interaction, such as PC-based instant messaging, voice calls, and face-to-face interaction. The voice call has obvious advantages (and dangers) in terms of developing mutual engagement. More tightly interwoven turn taking, mutual rhythm, use of tonal range, and use of pitch help a couple develop cohesion (Collins 2004a, pp. 48). If these elements are handled well, they can further encourage the relationship. If one partner dominates the conversation, or if one partner is halting or awkward (in other words, if the couple experiences a failed ritual), it is likely to be a damper on the development of the relationship (Andersen 2006). The effect of this process is to establish the mutual recognition of a common mood within the dyad. While the co-present interaction is a key to the establishment of a relationship, the mediation of information via other channels can also influence the establishment of a more serious relationship.

Once a relationship is established, mobile communication influences how the partners keep each other appraised of their status. Material from Norway shows that there is the exchange of endearments by lovers in the form of "good night" and "good morning" text messages. Examples range from the utilitarian "G'nite" (female, 15) or "Have a good night, hug" (male, 47) to the rakish "Do you want to spend the night? Hug" (female, 17) and "Good night sex bomb" (female, 35). Others are more mysterious: "Hug, has the prince slept well? Use the signal" (female, 48). Some are openly infatuated: "Nite, I love you" (female, 27). "I wish you a good night and I love you" (female, 28). Others refer to the internal lingo of the relationship: "Good night and sleep with an image of a sleeping bear" (female, 16). In the case of teens who still live with their parents, this represents a private communication channel where others are not available. This form of interaction is ignored only at great risk. Ito and Okabe (2005, p. 265) report many of the same types of interaction among Japanese teens. They note that for couples living apart "messaging became a means for experiencing a sense of private contact and co-presence with a loved one even in the face of parental regulatory efforts and their inability to share any private physical space." Consider these messages, reported by Ito and Okabe:

> I really want to see you (>_<). I am starting to feel bad again. My neck hurts and I feel like I am going to be sick (; _;). Urg.

> I get to see you tomorrow so I guess I just have to hang in there! (AoA).

These messages (by a young Japanese couple) include both text describing the emotional state of the individual and the characteristic Japanese emoticons. These comments are included in a longer, chant-like sequence of remarks. One partner was using the public transport system and the other was presumably at home. The ability to tuck these interactions into otherwise empty time is a unique characteristic of mobile communication. In this way the relationships becomes more omni-temporal. Texting allows the couple to maintain discreet, continual, and in some cases intense contact with one another. The interaction stretches from when the two part at school through the evening (including time spent doing homework, watching TV, and so on), interrupted only by bathing and sleep (ibid., p. 138). Indeed, teens in Japan, and also in Norway, have commented on the need to manage the communication flow so as to allow themselves time for private activities. Thus, although texting or using another social networking system is not a co-present activity, the partners are intensely focused and a common frame of mind is shared.

In language that approaches the notion of ritual induced cohesion, Ito (2004) comments that "this steady stream of text exchange, punctuated by voice calls and face-to-face meetings, defines a kind of 'tele-nesting' practice that young people engage in, where the personal medium of the mobile phone becomes the glue for cementing a space of shared intimacy." In the next chapter, I will examine this in terms of Licoppe's notion of connected presence.

In addition to facilitating remote interaction between partners, mobile communication helps to enhance co-present interaction. Ito and Okabe (2005) call this an "augmented flesh meet." They describe how two young people in Tokyo start the process of having a date by exchanging text messages during the hours before their actual meeting as they complete their work or their studies and as they negotiate the public transit system. After the date, they return to textual interaction as they go their separate ways via the Tokyo transport system and dwell on "the fading embers of conversation and contact." It is easy to see that the richest time is during the co-present phase of the date. However, the coming interaction can be planned and negotiated beforehand. Afterward, via mediated interaction, it can also be relived and reified as a part of their couple's lore (Ito and Okabe 2005, p. 271; see also Andersen 2006, p. 55). Were the battery to give out or were the connection to be lost because of spotty coverage, the interaction would be considered a failure.

A similar profile is seen in the exchange of text messages leading up to an hour-long telephone call on an evening when they will not see each other. This is then followed by a series of text messages in what Ito calls the "afterglow" of the telephone conversation. These text messages comment on the telephone conversation and perhaps embroider the various themes that were discussed in the telephone conversation. The teens use text messages to extend the face-to-face meeting or the telephone conversation. This does not make the co-present time less important; rather it may respectively foreshadow it and give them time to reflect on it from a slight remove. Like a cubist painting, it may expose the same event to multiple anticipations and interpretations.

The ability to extend the interaction beyond the period of co-presence (the actual point at which an individual "shows up") is also somewhat more diffuse. So long as there is interaction (mediated or co-present), the individuals are at least sharing time (Ito and Okabe 2005a). It is true that the real business of being a couple is best achieved when the two are together, but the preparation for the time together and the reflection on recently shared time are extended and blurred.

My discussion up to this point paints a rather puritanical picture of dating. There are obviously other more carnal forces afoot, and these are also being played out via the mobile phone. In research carried out in Norway, for example, there is a covariance between mobile phone use and sexual activity (Pedersen and Samuelsen 2003; Ling 2005).

The Norwegian researcher Lin Praitz has assembled a corpus of text messages that go beyond the rather coy discussions of flirting and finding partners into much more explicit territory. In this material, there is the expression of much clarity with regards a range of sensual desires (Proitz 2006). An example of this is the text message from an 18-year-old girl: "It's insane … Just met you three times, but have never felt anything like this before … Want so much to hold you … Kiss you … Going crazy!" In Proitz's messages we can read about teens who express the unbearable longing to be with paramours—if they can only separate themselves from current boyfriends or girlfriends. They reflect on their own sanity while apart, and they are lavishly frank in their description of their desires.

Bella Ellwood-Clayton, who studied texting in the Philippines, found many of the same forces at work. Texting has developed as an alternative to the more staid system of formal courtship. Consider this example (Ellwood-Clayton 2003, p. 233):

LETICIA: Does it mak u hapy 2 stel a kiss? Remember tho shalt not stel, best to ask!
CAPTAIN: I kno wen 2 do it
LETICIA: Com show me how so I myt also
CAPTAIN: Jaz lyk a magican … I nevr reveal a secret …
LETICIA: (sent a message with the graphic of a dancing bear)
CAPTAIN: wers my kiss?
LETICIA: Y dnt u cum n get it? Latr, im nt yet going 2 bed. I's stil her at d prayer meetn, prayn 4 you …
CAPTAIN: Ur not prayn. Ur thnkn of me.

This brings us back to the question of just where the ritual interaction is being carried out. The woman at the prayer meeting would seemingly fit nicely into the Durkheimian notion of focused co-present interaction. She should be drawn into the prayer meeting, sharing sidelong glances with other worshipers and joining in the common chanting and singing. She should be feeling the mutual effervescence of the co-present event. It should be such that the tempo of the sermon and the singing of hymns should draw her, along with the others who are physically present, into the flow of the event. Instead, it seems that the chief occupation of her thoughts and the chief form of cohesion being developed is associated with her eventual meeting with the Captain.

Ellwood-Clayton describes how the texting relationships exist in the gray zone between gabbing, entertainment, and earnest desire for a partner. The text relationships are in some cases only short-term flirtations; in other cases they solidify into actual (read: physical) relationships. There is, of course, the complexity of any affair of the heart where one partner is more serious than that of the other.

The work of Proitz and Ellwood-Clayton goes beyond the rather more innocent texting of Ito and the material from my analyses. In the context of ritual interaction, the material from Proitz and Ellwood-Clayton seems to include a lot of what we might consider sexual foreplay, albeit in a virtual mode. There is seemingly little question, however, that by the time that Praitz's couples met negotiations would be at an advanced stage.

In the rather dry language of ritual interaction, the social boundaries as to the in- and out-groups would have been well established, the effervescence of Durkheim and the common mood and entrainment of Collins would be well in hand, and there would be a high degree of mutual engagement in a small-scale Goffmanian ritual.

Conclusion

There are various situations in which the use of a mobile phone seemed to tear at the fiber of society. While the individual is perhaps well ensconced in a telephonic interaction, the effect on the local situation can mean that the co-present individuals have to make room for the actions of the telephonist. That is one side of the equation.

However, mediated interaction also plays into the ritual form, and thus it actually enhances the cohesion of the group, be it a dyad, a small group of teens (Ling 2005a), a church prayer group (Rafael 2003; Palen et al. 2001), or attendees at an Alcoholics Anonymous meeting (Campbell and Kelley 2006). I have examined situations such as greetings, or perhaps our willingness to dispense with them, the negotiation of romantic interaction, texting and the use of various forms of argot, humor, and repartee, and the use of gossip. I have outlined how the mobile phone affects how groups are constructed and how they are brought together in the interaction. The mobile phone is often adopted by pre-existing groups that were formed in other contexts. Thus, there is often a pre-existing solidarity or internal cohesion within the group. What does the introduction of the mobile phone mean in this context? The assertion here is that the way ritual is performed is carried over into the mediated interaction of the group. In some cases the interaction is modified and even strengthened. The lovers, the people telling jokes, the people exchanging gossip, and the people who use pre-configured greetings in general have a background that is established in co-present interaction. In these cases, however, the mobile phone has extended the range of the group interaction. The lovers can send "sweet nothings" to each other through the day and, in a small way, rekindle the mood of the last date. The friends or work colleagues can use humor in their interactions in order to get around difficult situations, to motivate the other person, or simply as a phatic device to let them know that they are valued. The friends can exchange gossip, either over the back fence or via the mobile phone, and in the process reaffirm moral standards of the group. Finally, pre-configured greeting sequences (or, in one of the cases cited here, picking up the thread of an existing interaction via the use of the mobile phone and the caller ID function) mean that the interlocutors maintain the same line of interaction as they carry out their alternatively separate and collective Saturday morning errands. The use of mobile communication influences the ongoing ritual interactions by giving these people another channel through which to interact.

In texting and in the use of jargon by teens, ritual interaction seems to be more uniquely related to mobile communication and the possibilities allowed by texting. The use of the z endings exists almost exclusively in the world of mobile phone texting. The use of faux Swedish also seems to exist there but not elsewhere. This points to the possibility of uniquely mediated forms of ritual. These formulations are obviously commented on when the teens are co-present. For example, they discuss how "out" it is to use the z formulations and how funny or positive it is to write the quasi-Swedish words. In this respect, the co-present interaction supports the mediated culture, and not the opposite (as in the case of romantic interaction).

Mobile mediated interaction has the potential to create group cohesion. As when Durkheim sees the group experiencing a mutual sense of themselves when they are "shouting the same cry, saying the same words, and performing the same action" (1995, pp. 231–232), the list can be extended to include joking, gossip, romance, and argot via the mobile phone. Arminen (2007) points to how mobile mediated ritual supports social cohesion:

> Chains of connotative meanings get established in this manner, and the coordination of social action intertwines with the symbolic organization of everyday life, establishing the actor's habitus that signifies the chosen way of life. In the ubiquitous mobile presence, the settings and activities may get symbolic embellishment. Seamless communication not only rationalizes time usage but intensifies social presence. The accountability of action extends both to the timing and social dimension of activities. Instead of accountability afterwards, the perpetual contact extends to the very moment of accomplishment of activity. Intimate connection to everyday life also enables social accountability of actions and choices through seamless communication. Depending on the social configuration, this extended accountability may strengthen external purely goal-rational control as well as social responsibility.

Like Licoppe (2004), Arminen is saying that the mobile phone extends the opportunities that we have to know each other. Connected presence is a construction that is contrasted with more traditional interaction between friends where there are relatively long periods of no contact punctuated by short "get-togethers." During the time where they are co-present the friends catch up on the main events in their respective lives. Licoppe contrasts this with connected presence that is carried out via the mobile phone. Since the threshold for interaction is lower, the members of a friendship circle can freely contact each other whenever and wherever the mood moves them. They can call or send text messages when the impulse strikes. This means that the "get-together" takes on a different tone. Since all are relatively updated due to the background traffic, the co-present sessions do not have the function of allowing the individuals to catch up. Rather, the "get-together" allows them to carry certain lines of narration further. Thus, the device is useful in extending the original bond beyond co-present situations into the folds of daily life. In slightly different terms, the interaction, both mediated and co-present, can be more effectively used to develop the ideology of the group.

The mobile phone extends the ritual reach of society beyond co-presence. Thus, the device extends the reach of parents, children, and friends. Rather than relying on a totemic representation of a social circle, they are, to the use phrase of Katz and Aakhus, in perpetual contact (Katz and Aalchus 2002a). The point here is that the mobile phone seems to result in stronger internal group bonds. This idea seems to fly in the face of the work of Putnam (1995), who fears for the reduction in social capital due to mediation. It also seems to contradict the suggestion by McPherson et al. (2006) that we have fewer and fewer confidants.

Section 5

The Dark Side

Online Communication and Negative Social Ties

By Gustavo S. Mesch and Ilan Talmud

The social network perspective focuses on exchanges (or the lack of) between pairs of actors. A social network relation denotes the type of exchange or interaction between any pair of actors in a system of actors. The network approach differs from other approaches mainly in its focus on exchanges and interactions between actors, not on the individual characteristics of the actors engaged in the exchange of resources. Social network analysis is used to describe the network and to explain how involvement in a particular network helps to explain its members' attitudes and behaviour. As most social networks reflect exchanges that provide companionship, social support, information and social identity, most analysis has naturally centred on the positive or socially accepted resources which are available through them.

However, social network analysis has also been used to study networks of deviant behaviour, such as drug use, criminal involvement and bullying. Most of this book has taken a social network perspective to understand the changes in social structure of young adolescents' social networks. From the perspective of social diversification we have discussed how online networks affect the size, composition, social similarity and range of contact of young adolescents. At the same time, we should note that since the use of information and communication technologies (ICT) began to expand there has been social concern with the potential negative effects of these networks, particularly regarding youth. In one famous case in the USA, the use of identity deception to harass a young female through a social networking site (SNS) resulted in the tragedy of suicide. Meigan Meier had been instant messaging (IM) with a 'cute' boy she had met over the social networking site MySpace. She struck up a friendship with 'Josh' and the pair exchanged dozens of messages. Josh was the fake name used by an adult female neighbour who lived in the same street. When 'Josh' told her 'he' didn't want to be her friend and called her a 'liar and slut' she became depressed. She committed suicide a few days later after her online friendship ended. In another case Ryan Patrick Halligan was bullied for months online. Classmates sent the 13-year-old boy instant messages calling him gay. He was threatened, taunted and insulted incessantly online. Eventually, in 2003, Ryan killed himself. These of course are extreme cases; online harassment for the most part has consequences, but not so radical. In this chapter we address this issue, the main task being to sort out what is new about cyberbullying from what is not and belongs to more traditional forms of peer harassment known before the advent of information society. The aims are to identify incidents of cyberbullying and harassment, to identify personal and family characteristics of youth who are the victims of such experiences, and to identify some consequences of such victimization.

Cyberbullying from a Social Network Perspective

In our understanding of the role of youth social networks it is positive outcomes that are almost always emphasized. From a normative social perspective, social networks have an important role on the individual and social levels. On the individual level, the existence of social ties is linked to the development of a positive social bond to social institutions. Having friends means developing social skills, as through social interaction we internalize social norms, expectations and social values. On the aggregate level, social ties link individuals to society and support the development of a common identity and social solidarity. Through social interaction, namely exposure to each other, individuals develop mutual obligations, trust and commitment. The existence of social ties affords the individual access to companionship, information and romantic involvement. In the literature on youth, social bonds with other peers have indeed been associated with compliance with social values and social norms, but with avoidance of deviant behaviour as well. Supportive and mutual ties have been linked to keeping clear of engagement in deviant and non-normative behaviour both offline and online, to avoid the negative consequences of suffering social sanctions at the hands of significant others.

At the same time we should recognize that social ties can carry negative outcomes, commonly thought to be the result of lack of social ties alone. Not belonging to a large network, not experiencing closeness to existing ties, or belonging to a low density network are all assumed conducive to deterioration in mental health (Beraman and Moody, 2004). Note, however, that negative outcomes may result from being involved in negative social ties—negative in the sense of hostile, aggressive and humiliating interactions (Beran and Li, 1995; Berson *et al.*, 2002). In this chapter we address these adverse aspects and focus on their prevalence, sources and consequences.

From Bullying to Cyberbullying

Bullying is a serious social issue as it is a vicious threat to a welcoming and supportive educational environment. Grave consequences have been identified in victimized people, including suicide attempts, eating disorders, running away from home, depression, dropping out of school and aggressive behaviour in adulthood (Borg, 1998; Kaltiala-Heino *et al.*, 1999; Olweus, 1999; Hawker and Boulton, 2000).

Historically, bullying is a common form of youth violence that affects children and teenagers, mostly when engaged in age-related activities such as going to school and travelling to or from school or when in public places such as hangouts (Patchin and Hinduja, 2006). Accordingly, in the past bullying has been extensively studied as a behaviour enacted in children's and adolescents' natural habitats such as the neighbourhood and school and at social gathering places. As to the prevalence of bullying, the data show that 10 to 15 per cent of students aged 12 to 18 years had been bullied in the previous 30 days (Devoe *et al.*, 2002; Galinsky and Salmond, 2002).

Bullying is conceived as an act of aggression with the attempt to exert domination through inducing fear. According to dominance theory (Hawley, 1999), students use aggression against weaker fellows to gain access to resources, including high sociometric status among peers. Bullies will therefore occupy more central network positions and hold more physical and social power, while victims will probably not be at the centre but more peripheral on the network than their classmates. Mouttapa *et al.* (2004) found that victims received fewer friendship nominations and occupied less central positions in the friendship network than other members. Socially, for different reasons, they were less integrated in the school groups and of an inferior social status in the school network.

An issue more relevant to our argument is the major importance of the youth network as a protective factor. During adolescence an attempt is made to attain social status in the peer network. Achieving good grades, participating in extracurricular activities, socializing and displaying cultural symbols are in part tools for adolescents striving to achieve social status. Targets of bullying at school tend to be children who are not well integrated in their social networks and known to be lonely or isolated. Well-integrated children are not targeted, as their friends are more likely to be in their company and to intervene in threatening situations (Mesch *et al.*, 2003). Network analysis has been used to investigate whether school sociometric status is associated with bullying victimization. Centrality, an index of popularity, is linked to both pro-social and anti-social behaviors. In their study of fourth and sixth grade males, Rodkin *et al.* (2000) found that boys perceived as non-aggressive, cooperative and leaders, and boys perceived as aggressive, equally occupied a central position in their classmates' social network. Some studies have specifically examined bullies' and victims' sociometric characteristics. Victims were often found to be rejected by their peers and lonelier than other students (Graham and Juvonen, 1998). There is evidence that a display of reciprocal friendships (e.g. students nominate a friend and receive a friendship nomination from that friend) protects students against victimization (Boulton *et al.*, 1999). Thus many young people are able to avoid the experience of bullying at school mainly due to peer or parental support (Farrington, 1993; Nansel *et al.,* 2001; Mesch *et al.,* 2003). Peers represent a support system, and studies on social networks and bullying have shown that victims are more likely to have friends who are non-aggressive, and offenders have ties to others who are aggressors (Mouttapa *et al.,* 2004). Another notable result of these studies is the importance of closeness of children to parents. When children inform their parents of bullying experiences, they are able to intervene and inform the school authorities (Farrington, 1993; Mesch *et al.,* 2003). These findings indicate the relevance of studying social networks for the understanding of bullying.

Concomitant with increased use of the internet has been increased reporting on cyber-harassment, sexual solicitation and cyberbullying (Berson *et al.,* 2002; Li, 2006; Patchin and Hinduja, 2006). Online bullying is an overt, intentional act of aggression against another person; it is wilful, intentional and repeated harm-doing, making rude or nasty comments about others online, spreading rumours and distributing short video clips that are offensive or embarrassing to the victim (Ybarra and Mitchell, 2004). Cyberbullying involves the use of information and communication technologies such as email, cell phone, text messages, instant messaging (IM), defamatory personal websites and defamatory online personal polling of websites to support deliberate, repeated and hostile behaviour by an individual or group—all intended to harm others (Rosen, 2007).

By utilizing information and communication technologies, bullying enjoys the advantage of several characteristics of the medium that transform the essence of the phenomenon as we know it. First, online communication in its very nature might induce bullying behaviour (Giuseppe and Galimberti, 2001). Communication that lacks nonverbal cues, status symbols and proximity to the victim may produce a behaviour that is self-oriented and not concerned with the feelings and opinions of others. Self-orientation may lead to lack of inhibition and negative perceptions of others, resulting in an increase in online bullying. Second, offenders exploit the internet's relative anonymity, through the use of screens or nicknames, to hide their true identity. An overall feeling of fear is generated in the victim, not knowing the perpetrator's identity; he or she does not know whether the perpetrator is or is not a classmate or a person met online (Li, 2007). Fear of unknown cyberbullies harms the educational and accepting environment essential to the classroom. The school is perceived as a hostile environment where victims feel unsafe, and such a child might avoid this by not attending classes. Third, the online environment provides a potentially large audience for the aggressive actions. This might appeal to perpetrators and furnish them positive feedback for their actions. Fourth, the large audience may amplify the negative effects of online bullying on the victim, as the harassment is being watched by all known acquaintances even beyond the school and neighbourhood. In sum, we may conclude that a salient difference between school and cyberspace is that in the latter a large number of perpetrators can be involved in the abuse, and classmates who eschew bullying at school will engage in it in cyberspace, hiding behind anonymity.

Past studies on real-life bullying have shown the importance of the audience, as 30 per cent of bystanders were found to express attitudes supporting the aggressors rather than the victims. The longer the bullying persists, the more bystanders join, and the more the bystanders join the worst are the consequences (Boulton *et al.*, 1999).

Prevalence and Consequences of Cyberbullying

An early survey in Canada showed that one quarter of young internet users reported they had experienced receiving messages containing hateful things about others (Mnet, 2005). Ybarra and Mitchell (2004) conducted a large study of young internet users in the USA and found that 19 per cent of the adolescents reported being bullied. Victims of online bullying were more likely than non-victims to be the target of offline bullying as well, but the correlation was far from perfect. A more recent online study of young internet users found that 29 per cent were victims of online bullying (Patchin and Hinduja, 2006). Online bullying seems to be increasing over the years. A study by the Crimes Against Children research centre that compared the results from two US national youth internet surveys in 2000 and 2005 found that self-reported online victimization of bullying increased from 6 to 9 per cent. Also, the percentage of children reporting cyberbullying others online increased from 14 to 28 per cent. In that study, being cyberbullied was defined as receiving mean, nasty messages, being threatened with bodily harm, being called names and having others tell lies about the victim on the internet. Inspecting in more detail the types of online bullying to which youth are exposed online, a study in the USA found that 13 per cent had experienced a situation of others spreading a rumour online about them, 6 per cent a situation in which someone posted an embarrassing picture of them online without their permission, and 13 per cent had received a threatening or aggressive email or text message (Pew Internet and American Life, 2006).

As to risk factors, studies have indicated that the higher the frequency of internet use, the higher the risk of cyberbullying (Patchin and Hinduja, 2006; Mitchell *et al.,* 2007; Rosen, 2007). Victimization occurs more often in internet spaces used for communication with unknown individuals such as chat rooms and social networking sites than through email and IM (Hinduja and Patchin, 2008). Recently a secondary analysis was conducted of the Pew Internet and American Life study of parents and teens (2006) on the link between teenagers' internet activities and victimization. When types of internet activities were divided into searching for information, entertainment (playing games online) and communication, it was found that only using the internet for communication was associated with the likelihood of cyberbullying. In particular, youth who had a profile in a social networking site or a clip-sharing site (such as YouTube) and participated in non-moderated chat rooms were at higher risk than youth who did not participate in these activities. Online games were not found associated with the risk of online bullying (Mesch, 2009b). An online profile in social networking and clip-sharing sites provides personal characteristics, disseminates contact information and exposes

Table 7.1 Summary of risk factors of online victimization

Online bullying	• An act of aggression, wilful and repeated harm, including making nasty comments, spreading rumours, sending offensive text messages and posting embarrassing clips or photographs.
Individual risk factor	• Victim of bullying at school. • Lack of supportive social ties.
Exposure factors	• High frequency of internet use. • Willingness to share personal and intimate information online.
Online activities	• Participation in chat rooms. • Participation on forums. • Having a profile in a social networking site.
Technology-induced	• Internet anonymity. • Reduced inhibition in making aggressive comments. • Large internet audience.

the adolescent to potential contact with motivated offenders, probably unknown. This private information is the raw material that might be used by potential offenders to call youngsters names, threaten them and ridicule them. Consistent with this approach, the same study showed that attitudes supporting disclosure of personal information predicted the likelihood of cyberbullying victimization. Youths who reported greater willingness to disclose personal information to other individuals when meeting for the first time were more likely to report being victims of cyberbullying (Mesch, 2009b).

On entering online space some adolescents are more willing than others to disclose personal information without being asked to. A recent study of young people's public profiles in MySpace showed that these contained personal information such as pictures of themselves with friends or family. A number of youth inserted pictures of themselves posing in swimsuits and underwear. Information on habits such as smoking and alcohol use could be found. Some even included their contact information such as the school they attended and phone numbers. Public disclosure of such information increases the risk of cyberbullying (Hinduja and Patchin, 2008). Not surprisingly, participation in chat rooms heightens the risk of cyberbullying still more, as participants are likely to engage in conversations with strangers, some of whom may be offenders. Studies have already found that online conversations tend to develop intimacy, and individuals are more likely to share private and personal information online because of the relative anonymity of the medium. Interestingly, playing online games was not found to be associated with risk of cyberbullying. Individuals engaged in this activity are most probably oriented to a less expressive and more instrumental form of communication, focused not on personal characteristics but on characteristics of the game (Mesch, 2009b). See Table 7.1.

The findings direct our attention to the role of online social networks, as online victimization is associated with the use of social media. This means internet applications that are used for communication, in particular those which expose youth to networks of individuals who are unknown or who are known but may use a false identity, facilitates their aggression.

Outcomes

For several reasons the effects of cyberbullying might be more pronounced than those of traditional bullying. An important characteristic of bullying is that when moving from physical to virtual space its intensity increases. In traditional bullying the possibility exists of physical separation between the aggressor and the victim, but in cyberbullying physical separation does not guarantee cessation of acts such as sending text messages and emails to the victim. Second, with the internet the abuser has a sense of anonymity and often believes that there is only a slim chance of detection of the misconduct. Third, when bullying is technologically supported, the aggressor is not aware of the consequences of the aggression. The screen does not allow a view of the victim's emotional expression. Anonymity and absence of interaction may make the aggressor still less inhibited and increase the frequency and power of cyberbullying (Heirman and Walrave, 2008).

Being a victim of online bullying has negative effects on adolescents' well-being. Victims are more likely to engage in high risk behaviours such as low school commitment, neglect of grades and alcohol and cigarette consumption (Finkelhor *et al.,* 2000). An emotional aspect also exists. After a bullying event the victim reports feeling angry and upset, and has difficulty concentrating on school- work. Online bullying has proven to have a negative effect on parent-child relationships and on relations with friends (Patchin and Hinduja, 2006). Regarding actions taken by victims, most respondents said they took none, 20 per cent decided to stay offline and only 19 per cent went to their parents and told them about the incident (Patchin and Hinduja, 2006). As these data show, the aggressors' perception that their acts incur no consequences seems to reflect the low likelihood that victims will take any action and report such incidents to teachers and parents. Thus, aggressors continue with their misdeeds because they do not face any negative social reaction.

Recently the danger has increased with the establishment of social network sites that provide platforms for publishing hate and cyberbullying. Examples are Enemybook, Snubster and Hatebook. These utilities describe themselves as antisocial; subscribers may add to the Friend feature, adopted from the basic idea of Facebook, an additional feature of enemies, the reasons for choosing them as enemies and spreading their secrets, and can invite others to join them in their bullying (Hammond, 2007).

Explaining Cyberbullying

The explanation of online victimization requires a socio-technical framework incorporating patterns of internet use, the nature of computer-mediated communication, and individual characteristics that affect the quality of ties. The pivotal element is the existence of opportunities for an aggressor. These are likely to arise as young people conduct their regular activities online: cyberbullying is possible only for adolescents who have access to the internet. That said, even for internet users the likelihood of being victims of cyberbullying varies according to the amount of daily time use, types of internet use and extent of disclosure of information. The concept of diversification is useful. We have shown in previous chapters that internet use for communication purposes is an important element in youngsters' social life. Participation in online communication is an important component in the explanation of cyberbullying. Consistent with our argument, frequent internet use and high level of internet skills increase the risk of being bullied online (Ybarra and Mitchell, 2004).

Participation through social media (email, instant messaging, social networking sites, forums and chat rooms) have an effect on the size, composition and quality of social ties, increasing exposure of youth to them. The effect varies according to each application, and the risk of exposure to victimization varies accordingly. Participation in chat rooms and forums and exchange of emails expose youth to new ties with unknown others. With some of them the exposure can have positive consequences as youth are able to meet others who share their interests and concerns and are able to exchange resources that are not located in the existing network. At the same time, internet activity enlarges the youth network; hence the risk to exposure to others with whom negative relationships may develop. The more unknown individuals become part of the social network, the greater the likelihood of meeting some who might become aggressors. The risk may also intensify due to the heterogeneity of the expanding network's composition. This raises the likelihood of the adolescent meeting others who are socially different (in age, race, ethnicity) from him or her, and social heterogeneity might be a further source of aggression. Some older individuals might express interest in associating with teens to harass them or solicit them sexually. Also, the addition of online ties to the youth network might temporarily decrease the strength of existing ties, reducing the sources of social support available to the individual. The extent that cyberbullying is carried out by strangers remains to be assessed as existing evidence is more anecdotal than based on large-scale studies. Due to the anonymity of the medium it is difficult to establish perpetrators' identities or if they belong to the victims' social network or not. As most of the adolescents' use the internet to connect with schoolmates and members of the peer group, it is plausible that the offenders are more likely to belong to the immediate social circle and less to be unknown strangers.

The use of social media such as IM to maintain existing ties and keeping a profile and communicating with friends through social networking sites expose youth to the risk of online victimization, but by a different mechanism. The incorporation of IM and social networking sites into adolescents' lives transforms their relationship with peers. When teens arrive home every day, rather than disconnecting from their friends they enter a state of perpetual connection as IM is on all the time, text messages arrive and conversations continue. In this case online bullying becomes an extension of school bullying, conducted after school hours through electronic communication. Now the possibility of after-school disconnection from or avoidance of contact with aggressors decreases. Cyberbullying often starts at the school or in the neighbourhood and it continues online, as affirmed by adolescents themselves. In one study, US teens were asked where they thought someone of their age was more likely to be bullied or harassed. It is interesting that while 29 per cent replied that it was more likely to happen online, the vast majority of these young internet users, 67 per cent, thought that it happened more often offline. Only 3 per cent thought that it happened equally online and offline (Pew Internet and American Life, 2006). For teenagers, online bullying can be emotionally debilitating, particularly because it is mainly a continuation at home of the aggression in the playground by known others from their social network. With traditional bullying, on arriving home the youngster felt safe and there was a respite from the aggression; now the aggression follows the adolescent home and can go on 24 hours a day. There is no place for the youngster to hide, even at home, from the persistent aggression (Rosen, 2007). A study in Canada (Li, 2007) investigated different forms of electronic bullying and found that the most frequently reported were email (20 per cent), chat rooms (33 per cent) and mobile phone texting (13 per cent). In the UK a study of the different channels of cyberbullying in greater depth found similar results. Phone calls, text messages and email bullying were the most common forms. Distributing a picture/video on websites and instant messaging were reported to a lesser degree (Smith *et al.*, 2006). The study also investigated the impact of each cyberbullying type and found that the picture/video clip was considered the most harmful by youth, followed by phone calls, text messages and website. Bullying by email, instant messaging and chat rooms was perceived as not very harmful (Smith *et al.*, 2006).

Important too is that online bullying requires some knowledge of the victim. When conducting online activities, individuals differ in their readiness to share personal information. Some are less willing to provide contact and personal information than others. Providing personal information can be considered a risk factor for victimization, particularly when it is given to strangers (Mesch, 2009b).

Another significant element is the way in which the internet is perceived; this affects the user's behaviour. In many ways the effect of the added value of communication technologies to face-to-face bullying is the well known disinhibition effect, often associated with the use of these technologies. Suler (2004) noted that cyberspace can be conducive to behaviours that may not be revealed in the offline world. Various factors contribute to creating a subjective perception that fosters the disinhibition effect. Among the most salient characteristics are anonymity, asynchronicity and dissociative imagination. Anonymity means the individual's subjective perception that the use of a nickname online separates him or her from the real world and his or her real identity is not known to others. This sense of anonymity might give rise to actions for which the actor does not feel responsible, at least not in the way he or she feels responsible for actions performed in a social circle in which his or her identity is known to friends, teachers and parents. Asynchronicity means that often individuals interact online not in real time, especially when communicating through email, forums and social networking sites. Not interacting in real time means that individuals do not experience in real time the reactions of others, or get immediate negative feedback for their actions. This lack of interactivity might multiply aggressive acts against others as their reactions and feelings are not revealed immediately or as part of the social interaction between the teens.

Dissociative imagination means a situation in which some individuals evaluate online events differently and separately from face-to-face events. Certainly anonymity may reinforce this feeling that online norms of behaviour unacceptable in a face-to-face situation are acceptable online. This is sensed as a different sphere, less real, or with fewer real consequences, for the victims. Disinhibition can be linked to the 'lack of social cues' perspective, which holds that the deficit of online communication lies in the lack of social cues, of nonverbal signals that denote social status, and of the emotions of the participants in interpersonal communication. Accordingly, in the case of cyberbullying the victim's emotional reactions are not present. The aggressor is unaware of the victim's distress, fear, tears and other emotional reactions. He or she is thus likely to increase the aggression attempts without setting limits on it that would otherwise derive from interpersonal contact and interactivity. These perceived characteristics cannot be taken as the sole factors responsible for all online bullying activity. They work in combination with the aggressor's personality and family characteristics, which they interact with to amplify their effect.

Linking Traditional Bullying and Cyberbullying

Is cyberbullying independent of bullying at school and in the neighbourhood? Some observers have advanced the idea that cyberbullying is a new phenomenon, largely the result of children and adolescents coming into contact with strangers and unknown individuals in open chat rooms online. In these spaces of conversation and social interaction strangers hold conversations with young people, making derogatory ethnic and sexual comments, or deriding their contributions. However, recent studies point more and more to a link between school bullying and cyberbullying. A study in Canada found that the most important predictor of cyberbullying victimization was victims being bullied at school (Li, 2007).

Bullying and cyberbullying are closely related, and in many cases the bullying possibly started at school and/or in the neighbourhood and then spread to cyberspace. In fact, cyberspace serves bullies as yet another venue in which to harass others, as they take advantage of the high rate of internet use among youth. Another possibility is that bullying starts online, and later the perpetrators take it into the real world, converting it into face-to-face bullying.

In trying to resolve this issue we face the problem of identifying the aggressor. As the internet often provides anonymity, victims do not always know who the perpetrator of the aggressive behaviour is. A study reported that 25.6 per cent of the respondents said that they were cyberbullied by schoolmates, 12.8 per cent by people outside school. The most surprising finding was that 46.6 per cent did not know who cyberbullied them (Li, 2006).

Over time, online bullying seems to be on the rise. The percentage of victims of cyberbullying has increased despite the decrease in participation in open chat rooms and forums and an increase in the participation in IM and social networking sites. These results provide an additional indication that cyberbullying is aggression by known others, such as schoolmates, taking advantage of social networking media (e.g. Instant Messenger, social network sites, SMS). As noted earlier, a study by the Crimes against Children research centre compared the results of two national youth internet safety surveys in 2000 and 2005 and found that the amount of cyberbullying had increased. In Rosen's study (2007), although only 11 per cent of the teens reported being harassed, 57 per cent of parents and 34 per cent of teens reported that they were concerned about harassment on MySpace. Clearly, the use of communication channels that link youth with friends and friends of friends is associated with an increase of cyberbullying; this supports the argument of a link between this and offline bullying. Rather than being new, cyberbullying seems to be a supplement to school aggression.

Online Harassment

We have already argued that the internet is a space of social activity. Youth activities include participating in online games, online discussions, online social support and social networks. Participation in this social space might expose youth to the risk of online harassment. This notion refers to unwelcome and uninvited comments or attention. The comments provoke negative emotions and are insulting because of their gender or ethnic content. This can take the form of offensive sexual or racist messages, jokes and remarks purposely initiated by the harasser to humiliate the victim. Youth are victims of this type of online harassment. A study in the USA reported that 62 per cent of youth participating in a representative national survey reported receiving unwanted sex-related emails (Mitchell *et al.*, 2007).

A form of harassment that creates public concern is online sexual solicitation. Several large-scale studies have been conducted to assess the prevalence of online sexual solicitation. Regarding the frequency of harassment, a US study found that 13 per cent of youths aged 10 to 17 years reported experiencing an unwanted sexual solicitation on the internet in the previous year (Wolak *et al.*, 2003). In the UK, 9 per cent of children and youth aged 9 to 17 reported having received unsolicited sexual material online and 7 per cent reported receiving sexual comments online (Livingstone and Bober, 2004). In Canada, a study asked about the frequency and place of sexual harassment in the previous year, providing us with a rare opportunity to compare the frequency of sexual harassment in different social contexts. According to the data, overall 12 per cent of adolescents in grades 7 to 11 reported having experienced sexual harassment in the previous year. There was a significant gender difference. Nine per cent of the boys and 14 per cent of the girls reported sexual harassment. When asked about the space in which the harassment occurred, 8 per cent said on the internet, 6 per cent at school, 2 per cent on the phone and 2 per cent on the cell phone. The study also asked about the relationship with the offender and found that 52 per cent reported that the harasser was known from the real world while 48 per cent did not know who the harasser was (Wing, 2005). The results are consistent in other countries as well. A study conducted in Australia asked 502 young children aged 8 to 13 years about the extent they were exposed to different experiences; it found that 5 per cent were exposed to obscene language (Kidsonline@home, 2005).

The few studies that investigated online sexual solicitation indicate that the likelihood of being a victim of this aggression depends on the application that teenagers use. One study found that the locations where youth most frequently reported online sexual solicitation were email, social networking sites and chat rooms (Ybarra and Mitchell, 2008). But blogging in general was not related to sexual solicitation. Bloggers who customarily interacted with unknown others were the only ones found to be at risk of sexual solicitation (Ybarra and Mitchell, 2008). Adolescents who blogged proved more likely than youth who did not blog to post personal information online, including their real names, age and pictures of themselves, as well as disclosing personal experiences. Yet as the study shows, blogging in itself is not related to increased risk of online sexual solicitation. The risk of sexual solicitation becomes high only when the teen interacts with people met online. Furthermore, most of the youths who write a blog do not interact with people they meet online and do not respond to their messages (Ybarra and Mitchell, 2008).

Adolescents' choice of internet applications probably reflects a selectivity effect. Youngsters who have unsatisfactory ties with parents and friends choose to use chat rooms to compensate for this by engaging in communication with unknown others. Such activity is a risk factor for online sexual solicitation. For example, a study that compared adolescents who used chat rooms with those who did not found that chat room users were more likely to have low self-esteem, to not feel safe at school and to have been physically abused in the past. Thus, the study reported that for boys and girls alike chat room use was significantly associated with adverse psychological characteristics. In contrast, other internet activities did not show a consistent pattern of positive associations with these factors (Beebe et al., 2004).

Racial and ethnic online harassment has been reported in studies conducted in chat rooms in which adolescents participate. The anonymity of online interactions may lower control of racist remarks. A study that compared moderated and non-moderated adolescents' chat rooms reported that in the latter an individual had a 59 per cent chance of being exposed to racial and ethnic harassment remarks, and in the former a 19 per cent chance (Subrahmanyam and Greenfield, 2008).

The aggressor in this type of online harassment is usually an unknown other online user who was met briefly in a forum or chat room; he or she identified the intended victim's social status characteristics and set about harassing them. A follow-up study that investigated risk factors of online harassment found that the likelihood of exposure to sexual solicitation and sexual harassment was associated with exposure variables: high frequency of internet use and high frequency of participation in risky sites. Age was also related, as young adolescents were more likely than older ones to undergo this experience (Fleming et al., 2006).

In sum, bullying and harassment have not just moved from physical to virtual space, their intensity has magnified as well. Physical separation of aggressor and victim does not guarantee disengagement and cessation of acts of bullying—not in terms of frequency, scope, or severity of the inflicted harm. With the advent of Web 2.0, cyberbullying includes the use of email, chat, instant messaging, clips and blogs, which serve to embarrass and threaten, to make rude or vicious comments, and to spread rumours or clips and photographs of the victim in embarrassing situations. Cyberbullying, a serious form of cyber-harassment, has a number of important components:

1. Cyberbullies are aggressors who seek implicit or explicit pleasure through the mistreatment of other individuals.
2. Cyberbullying (like bullying) involves repetitious harmful behaviour.
3. A power differential between bullies and victims should be expected, and in the case of the electronic environment this differential might also be observed in computer literacy.

Internet users often believe that there is only a slim chance of misconduct being detected online, so threats and harassment have become prevalent among young users (Lanning, 1998).

Inquiry into young people's social networks today requires study of the patterns of internet adoption and of online and face-to-face networks. Adolescents' adoption and use of the internet are related to the social network to which each of them belongs. The choice of internet social applications such as forums, chat rooms, email, instant messaging and social networking sites depends on which online activities are carried out by others who belong to the peer group. A member's engagement in online communication requires that the other members have adopted the internet for communication purposes.

Internet adoption and integration into routine communication with others has an effect on a youth's access to positive and supportive ties, but also on the extent of his or her exposure to negative encounters and persistent harassment, with all its negative consequences to his or her well-being. From the studies reviewed, it is clear that some adolescents become exposed to repetitive aggression and harassment. Such exposure depends upon the online activities being conducted. The motivation to use forums and chat rooms is undoubtedly different from the motivation to use instant messaging and social networking sites. From the psychology literature we learn that shy individuals with higher levels of social anxiety are more likely to use forums and chat rooms. From the sociological literature we learn that individuals wishing to expand their networks and access to others who share their interests and concerns are more likely to do so. Through these applications youth are better able to establish social ties with others who share their interests and hobbies; but at the same time they become exposed to strangers who may belong to a socially different group, hence to some risk of harassment by strangers.

The motivations to use instant messaging and social networking sites are apparently different, having more to do with an attempt to maintain and reinforce existing social ties with present friends met at school and in the neighbourhood. The greatest danger of cyberbullying thus seems to be less from strangers than from known others; and as noted, cyberbullying is less likely to be a new behaviour than aggression which has acquired an additional medium—cyberspace. The use of the internet accordingly amplifies negative behaviour, allowing it to be conducted at school and after school hours as well.

These conclusions should be qualified, as bullying is a relatively new and developing behaviour that has to be monitored over time. As new applications are being developed, including social networking sites which openly instigate calling friends names and identifying them as enemies, we must follow the progress of cyberbullying and its connection with face-to-face and online behaviours. What is new about cyberbullying seems to be diversification of sources of risk (online activities) and diversification of kinds of bullying (using a new medium to bully).

Putting Social Context into Text

The Semiotics of E-mail Interaction

By Daniel A. Menchik and Xiaoli Tian

A set of studies of computer-mediated communication focuses primarily on inter-actants' impressions of e-mail as a medium. Many use the framework of social presence theory, which implies that problems in e-mail arise because it provides a low sense of awareness of an interaction partner (e.g., Short, Williams, and Christie 1976; Rice 1993; Rourke et al. 1999). Others examine the level of "information richness" permitted in e-mail and in face-to-face communication (e.g., Daft and Lengel 1984; Trevino, Webster, and Stein 2000). Social presence theory compares mediums in terms of the level of warmth and "personalness" they support; the information richness approach examines the level of personalization they allow and the number of senses usually involved in interaction (Rice 1992). Both argue that the reduction of contextual, visual, and aural markers in email results in a general drop in the quality of interaction (Culnan and Markus 1987).

A body of social psychological communications research explicitly addresses the *content* of online interaction in e-mail discussion groups, and this research often emphasizes users' difficulties. Multiple studies report complaints about the limitations imposed by the medium itself (Conner 1992; McCarty 1992). Some argue that e-mail discussion groups are not well suited to discussing or solving intellectual controversies among researchers, as individuals rapidly become unsatisfied with the contributions of others and are reluctant to commit themselves to interaction (Hiltz 1984; Harasim and Winkelmans 1990; Lewenstein 1995). E-mail discussion group participants' concerns over the low quality and unfocused nature of discussion frequently lead them to attribute limited value to the online context for in-depth interaction.

Sproull and Kiesler (1986, 1991) suggest an underlying reason for the perceived reduction in quality, arguing that problems in e-mail result from its inability to compel the user to limit the range of subjects and comments she considers appropriate to discuss. Their influential "social context cues theory" is based on the assertion that computer-mediated communication lacks equivalent cues to those available in face-to-face contexts, such as facial expressions, body language, and tone of voice (Kiesler and Sproull 1992). This leads to what they refer to as unregulated behavior in e-mail, which they argue is responsible for the increased number of misunderstandings, extreme reactions, and "irresponsible" activities that occurred in the e-mail discussion group used in an organization they studied (Sproull and Kiesler 1986, 1991; Kiesler and Sproull 1992). Other researchers have applied their theory to explain why aggressive and hostile exchanges between communication partners seem increased and why the usual inhibitions that govern interactions with superiors appear lower in e-mail (e.g., Dubrovsky, Kiesler, and Sethna 1991; Lea and Spears 1992; Walsh and Bayma 1996; Cramton 2001). They support Sproull and Kiesler's argument that the lack of face-to-face cues in e-mail creates a psychological state in which social and normative influences have been undermined.

In sum, communication theorists emphasize e-mail's negative effect on interaction quality and assign responsibility to the fact that the communication partner is absent. Researchers make comparisons among mediums by classifying them in terms of level of warmth exchanged, number of senses involved, or number of cues permitted. E-mail is thought to allow less warmth and to employ fewer senses and cues. Social context cues theory claims that these limitations undermine social and normative influences, producing "unregulated" exchanges that would not occur in a face-to-face context.

Since we are often able to communicate successfully online *despite* the change in context, sociologists must study how this is accomplished. Below we address this problem by presenting our observations of successful and unsuccessful attempts, evidence from participants on their intentions for the interpretation of their messages, and a theoretical framework that interprets how text organizes these social interactions.

Data and Methods

The first author studied the research organization that coordinated the assembly of panel members from the time of its development in July 2003 until its close in February 2005. Since the organization has a strong international reputation among social scientists and those engaged in media-related activism, the request for applications was widely diffused through well-populated e-mail discussion groups and the organization's website (which receives over 10,000 visits per day). It also invited specific well-known scholars and activists to apply for participation in the group. The 12 who were selected in the three-round process represented organizations and universities of various sizes from six continents. Two-thirds of the participants were native English speakers, and all were fluent.[1] E-mail was fully available to and used frequently by all.

The data from the project were collected from environments unmodified for the purposes of this study. Much research on e-mail occurs in controlled settings (although see Kraut et al. 1998; Kendall 2002), which is valuable for theory testing as it allows complete control over the conditions and characteristics of the groups being investigated (e.g., Yamigishi 1995; Kollock 1998).[2] Analyzing behavior from social groups in natural settings, however, allows for a longer time frame and a richer environment than is often allowed in the laboratory (Hedstrom and Swedberg 1998; Anthony 2005).

Multiple types of data were collected. First, every e-mail ($N = 338$) was read and double-blind coded according to variation along several dimensions: open-endedness (question or comment), subject (related directly to panel topic or not), whether it received a reply (as indicated in the subject field), intended recipient (if possible to determine), and whether it catalyzed or was formulated in response to a problem. We inductively developed categories of problems and respective solutions from these observations and participants' responses to interviews and open-ended survey questions.[3] After noticing problems in the e-mail discussions around senders' intentions for message interpretation, we contacted individuals via e-mail to discuss specific online activity.[4] As most participants reported that they were personally unfamiliar with each other prior to the collaboration, the e-mail discussion group appears to have been the primary medium for initial and subsequent interaction.[5] Second, two online surveys were administered to collect findings on individual sentiments regarding the project after approximately 5 and 15 months of duration (the average response rate was 90%).[6] Third, a 20-hour midproject meeting with all of the participants was observed by the first author and the proceedings were recorded and transcribed.[7] Fourth, semistructured in-person interviews were conducted with 10 of the participants to collect information about their experiences with the project.

The population we study is unusual in its openness to e-mail as a medium of communication and its capacity to use it; participants frequently employ sophisticated internet-related tools to organize meetings and other events.[8] Our study of how the system works under these ideal conditions identifies constraints and develops principles for explaining interaction in other online contexts.

The Project's Trajectory

Soon after the panel members were chosen and the e-mail discussion group was established, members began requesting feedback on subjects germane to the project. One participant asked how others measured the use of internet resources in public computing centers. Another wondered whether cell phones would be superior to computers for mobilizing their constituencies. A third suggested that it made less sense to purchase computers than to strengthen existing capacities of the nongovernmental organizations that work with movement leaders.

Responses rarely followed these posts, and enthusiasm for the project began to wane. The large number of unreturned messages led the project director to express his dissatisfaction with the progress of the e-mail discussion group. His sentiments immediately drew a series of responses in which members claimed that an offline meeting would catalyze future posts online. See figure 1 for a description of the volume and type of messages for the duration of the project.

The meeting occurred three months after the project began. Participants came to the meeting optimistic about its value in sparking substantive interaction around common interests. Their confidence was not misplaced; in the face-to-face meeting they seamlessly discussed subjects ignored online. Individuals contributed expertise and their own experiences in response to others' ideas and freely challenged perspectives. They needed no moderator, and subgroup exchanges continued through the lunch break. Participants interviewed at this meeting said it improved their trust and respect of others in a way they thought difficult to accomplish online but necessary for lubricating future e-mail interactions.

Despite the success of the meeting in supporting interaction around subjects initially broached in the e-mail discussion group, it only minimally catalyzed e-mail discussion (see fig. 1). Findings from a survey conducted the following February demonstrate that participants' reported desire to respond to e-mails had not changed relative to feelings expressed prior to the meeting, and in the following three months, the quantity of messages *declined* by approximately 25%. The expectation of a positive effect of face-to-face interaction upon sustained e-mail interaction was mistaken. To understand the origin of the group's difficulties, we must examine their primary problems and solutions.

The Problems and Solutions

A number of distinctions between offline and online interactions were problematic for participants or consistently appeared to present difficulties from an observer's viewpoint. These problems were useful in that they often provided clues for identifying unique characteristics of interaction in the online context. Three primary dynamics influencing attempts to discuss similar subjects across contexts were inductively exposed: *problems with terminology, problems of relevance,* and *problems with situational and background ambiguity.* In this section we present how these dynamics were experienced, as well as how participants adjusted to each of them.

Terminological problems.—Members of the interdisciplinary group repeatedly encountered and discussed terminological ambiguity. A sequence of three exchanges effectively demonstrates how each context offered different options for specifying the meaning of a sentence or its components, in addition to different capabilities for negotiating problems of meaning.

In an early e-mail, a participant called attention to the definition of a term frequently used by group members.[9]

>I am curious when we use this term 'civil society' if it is overly generalising a specific set of actors.

Let me quickly illustrate this using the example of 'government' as an actor: often the Treasury in the UK has opposing views from the Home Office,

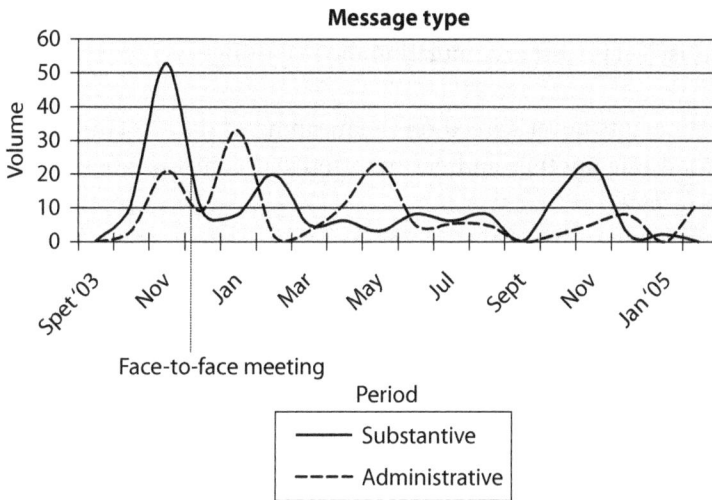

Fig. 1.—Volume of e-mail messages over the duration of the project, by type

who are often opposed in the Commons by backbenchers, with differing opinions presented by the House of Lords, etc. So I am always hesitant to use the term 'and the UK Government wants …'

So when we speak of consulting with 'civil society', or 'giving civil society a seat at the table', or 'civil society organisations are calling for …', I wonder if we are being overly generous to these specific organisations, giving them a larger hat than they deserve; or perhaps we are lacking analytical rigour and not looking specifically at what these specific groups want.

That's really what my point was; I am not calling for some definition of what an NGO is; I'm calling for scrutiny of blanket statements to see what the mess beneath is that is being concealed.

This e-mail attracted no responses. Indeed, the entire thread of discussion around civil society ended.

Such linguistic clarification of key words proceeded differently in the subsequent offline meeting.

Chi-Chen: First we must discuss the dichotomies we keep using. For instance, Asia's place in a so-called North/South divide is not clear.

Victor: And also, a shared physical presence does not always mean a shared point of view. We just assume this to be true.

Cindy: This language is not a geographical division nor an ideological division. In many cases, northern NGOs identify with the global South. Yet there are different problems in Asia than there are in Africa.

Pat: North/South, developed/underdeveloped, poverty/richness are all frameworks for new complex situations.

Gustavo: All classifications and language are loaded with power disparities. In this case, isn't capitalist accumulation the real thing?

Although participants never agreed on the meaning of the "North/South" dichotomy, they understood one another and easily negotiated concerns about ambiguity when meeting face-to-face.

But what occurred when panel members renewed their efforts at substantive collaboration in e-mail? Here we present an exchange in which they briefly experimented with communication among a smaller group of four individuals as a response to their premeeting difficulties. Each had written a summary of their work, and they collectively sought an approach to combining these into a larger piece to ultimately publish on the sponsoring organization's website.

>I am not sure this works too well and coherently. I don't intend to sound didactic but I think it would appear plagiaristic to reproduce large amounts of text of an already written and published piece—even if only published online—without clearly acknowledging the earlier publication or version … Pat

Victor replied to Pat with anger and indignation over his tone and word selection.[10]

>Plagiarist? I am not sure how you can use this term when we are expected to be joining papers together? Wasn't that the purpose of this group effort? Also I am only referencing material that you have listed in your paper. In my books plagiarist is a term used for when people copy others people work and give them no credit for it. Which in this case is not the case, as you are not the original thinker in a lot of the material and your name is listed on the paper itself!

I propose that we talk on the telephone to discuss next steps forward. …
Let me know a good time and numbers to call. Or whether just to call it off.

Pat's subsequent response to Victor exposes his frustration with being unable to communicate the meaning he associated with his words. He specifically draws attention to the importance he places upon the use of quotation marks.

>Here is my response and correction on the sensitive issue of plagiarism. I knew this term COULD be offensive and I tried, obviously unsuccessfully, to carefully choose my words. I wrote "I think it would appear plagiaristic" (though I failed to put the word 'plagiaristic' in quotes as I had intended). A careful reading of this, I believe, is that I am not absolutely certain and I am not calling Victor a plagiarist as he seems to imply, rather I am directing the possible 'appearance' of plagiarism to other (third party) readers—at [the research organization] or visitors to its website.

It seems I assumed too much. I was taught at my university that borrowing material substantially from one piece of essay or work and reusing it in another (even though it is still the same person's work) is grouped under plagiarism penalties. In other words, in this context I too would be party to any possible perception of reproducing my already submitted work and re-submitting it; and not acknowledging its earlier publication would foreground this 'appearance of plagiarism' even more. So this was as much an introduction of a point of discussion as it was a pointer to possible 'dangers'. That's the long explanation of it; whether it clears the misunderstanding I leave it to you.

Victor amiably accepted Pat's explanation.

>I am hoping that you can add to the present piece either plagiarising (kidding :-)) or adding a new piece.

Pat's renewed confidence in the group's capacity to collaborate indicates that the conflict was ameliorated by Victor's extension of goodwill.

>Thanks for the reconciliatory tone. I favour bringing to the joint paper a different piece—along the lines of Cindy and Stefan's contributions—and I can do this in the next few days, hopefully before the weekend. I am hoping that you will be willing to work on the joint piece again; I will try to help with this. At the end, we will remember it's been a long and torturous journey.

The final three messages here contained two frequently used forms of adjustment. First, Pat used capital letters and quotation marks to place special emphasis on particular points. He felt that this allowed him to insert volume and tone into a message communicated via a medium that disallows auditory emphases. Second, Victor sought to annul the prior meaning of the controversial word by accompanying it with an emoticon.[11] Pat recognized Victor's intention, noting the message's "reconciliatory tone." Although interpretative difficulties between two individuals are usually not so aggressively negotiated, this exchange exposes how the absence of face-to-face cues can influence the way language is interpreted among those who otherwise share much in common and hold a strong interest in working together.[12]

Taken together, these episodes reveal how approaches to managing terminological ambiguity differed across contexts and carried different consequences. The first e-mail demonstrates the difficulty of discussing conflicting meanings online, whereas an analogous offline discussion drew lively interaction. Pat and Victor's e-mail exchange demonstrates how initially divergent interpretations of the purpose and meaning of a word's use can damage relations, as well as the way punctuation and capital letters were thought to be able to clarify senders' intentions for how their messages should be received.

Ambiguities around relevance.—Several of the discussion group's problems stemmed from the disruption of the turn-taking format that characterizes face-to-face conversation (Sacks, Schegloff, and Jefferson 1974). Consequently, it was difficult for individuals to signal to whom their comments were oriented. These problems were indicated in both interviews and reflections expressed in e-mail.

Interviewees repeatedly pointed out problems arising from the asynchrony of message transmission and reply. First, message senders were unsure that others would recognize how their comments were germane to the points they sought to address. Since individuals were not online simultaneously, another member might broach a different subject before the designated recipient of an original post could respond. The resulting crisscrossed threads of conversation increased people's reluctance to post. Second, individuals were frequently unsure of the person or subject to which a point was oriented. People remained silent, despite having substantive and relevant ideas to contribute, as they later asserted, because the eye contact available in offline conversation for designating an intended respondent was unavailable. Participants felt little reassurance that they could orient others' contributions on a topic as they could have in face-to-face exchanges.

One sign of members' powerlessness to establish common subject matter was their feeling that the group did not share enough in common, despite the fact that many of them had researched the same subjects and published in the same outlets. Participants' responses to open-ended survey questions indicate that they felt like their ideas on a subject were not worth presenting: "I only post if have something important to say"; "The tone of the list-serv was poorly set. It was created saying: 'Talk amongst yourselves,' and that's all. I'm not too good at that"; "I don't know what we are supposed to talk about here. ... We were just thrown together to interact."

This perception was further evidenced when participants expressed a need for a shared goal: "Sometimes inventing even a simple goal—e.g. drafting a mission statement or selecting our next meeting place—can get things going"; "The list was a forum, but it didn't have a goal-oriented task, so I never felt an urgency to participate. Other, more urgent tasks crowded it out in my attention"; "You need a specific vibe or purpose for exchange to work, the issue of 'civil society and ict [information and communication technology]' didn't provide that in this case."

As mentioned above, participants expressed a desire to have a face-to-face meeting after the program director inquired into the low volume of messages. This request was particularly motivated by their belief that a meeting would offer a context facilitating a more focused discussion around common interests.

>I do believe that personal face-to-face contacts still make a difference, in creating and making active something like a netwroks which is bound to work long-distance. I felt positive about this opportunity and would be honoured to develop ways to share ideas and visions.
Louis

>And yes, the face to face is required to start things off.
Gustavo

>I think this network will benefit greatly from a face-to-face meeting. … For me at least, it will make this activity "real."
Jose

At the meeting, people spoke frequently about many of the topics addressed—and ignored—online. They did not each necessarily respond to the person who preceded them, but rather waited their turn to talk and to claim their share of the attention in the room. The diversity of their backgrounds and lack of a common goal appeared benign.

When they returned to online exchanges, however, a set of techniques emerged that reflected a more relational orientation. First, the author of a message would name the author of the previous one in the body of her message:

>But I also agree with Pat that, together with substance and isssues raised, we should also become more "operative" and concrete. I think in our restricted network meeting we did have some good ideas and I, starting from there, am also throwing a couple of ideas in the air.
>In response to Roger's message today, I venture to share the following intuitive thoughts …
> Following from the 'brainstorming type of exercise' recommended by Chi-Chen, I'll try to weigh in. I have to admit that a part of me was still recovering from [the meetings] even when I returned from the holidays, only to start teaching again, but here's my effort.

The informal pressure applied by these "linking" techniques improved substantive continuity in discussions. The referenced persons would respond, sparking a chain of interaction arising out of an initial effort to assert relevance. Interestingly, the response time dropped for messages with these characteristics.

Formally addressing a message to a specific individual similarly improved the likelihood of a response.[13] Exchanges between two people on project-related subjects were more likely to be sustained than those either without a greeting or with a generic one such as "Dear all" or "Dear colleagues." When participants with messages that *were* intended for the entire group addressed the project director or the participant whose comment preceded theirs, it appeared to others that the comment was situated within a dyadic interaction rather than framed as a comment or query to the collective. Even then, people were more likely to insert their opinion than if a dyadic exchange had not yet commenced.

Second, people would use others' posts to support their own contributions, choosing to cut and paste from previous e-mails to ensure that the intent of their comments was clearly indicated, as in the following message:

>A very late response … but possibly in this case not too late.

>>Let me take advantage of this first communication to throw an idea out there. It seems many of us are doing work that explores specific examples of uses of ICTs by CSOs [civil society organizations]—from content management and open source to wireless PDAs [personal data assistants]. … Does it make sense to put to together a casebook that can be used by CSOs to learn about the advantages of "cutting edge" technologies as well as how best to set themselves up to exploit them?

>I have discussed this with my colleague Jane who thinks it can be very useful.

As people occasionally objected to the recontextualization of their words, this approach did not necessarily resolve problems of relevance. However, it appeared to be a tool through which members felt they could signal the temporal and topical referents of their posts.

In sum, project members initially thought that e-mail was a poor medium for substantive discussion because of both the characteristics of the medium and their own divergent backgrounds. Offline, this diversity was not a problem. Later in the project, relational techniques for signaling the situational relevance of a post were developed and frequently deployed.

Obscured situational and background information.—Finally, participants' inability to draw upon certain information from others' appearances and inflections provided another set of frustrations. The distance made it especially difficult to convey the intention behind an utterance.

In open-ended survey responses, the members of the panel indicated that they were initially reluctant to send e-mail because of uncertainty over whether they could calibrate their post to be received in a certain way: "I don't know how people are going to react"; "[In e-mail, you] don't know who you are talking to."

They were similarly uncertain about whether others would understand the tenor they sought to assign to a message: "[I am concerned about misinterpretation] if the e-mail involves some complicated personal problems, feelings, emotions"; "Conflict was problematic. I'm concerned that people would classify this kind of interaction as severe. Would be seen as flaming."[14]

Participants' concern over being perceived as aggressive was further revealed in their increased tendency to preface comments with disclaimers:

>I don't intend to sound didactic, but …
>This may come across harshly, but …

Interview and survey responses revealed an acute consciousness of the range of ways a message could be received, and the sender's inability to adjust accordingly: "I don't know if there are common understandings of the message. People essentialize the meaning of comments"; "The costs of developing your message in such a way that you can speak to everyone is too high." Rather than risk misinterpretation and the accompanying anxiety, participants often opted for silence.

Participants gradually developed ways to include certain information on their cultural background and emotional state in order to influence others' interpretations of their utterances. For instance, they would include details on their frame of mind at the time of posting, closing e-mails with disclosures related to their potential levels of awareness (e.g., "writing at 7 a.m."). Second, in the openings and closings of a message they might use foreign languages to signal an affiliation with a particular culture ("ciao," "colegas"); they might also index their culture in a celebratory gesture ("SELAMAT TAHUN BARU!" "Feliz Afio Nuevo"). Finally, they would often place information on their geographic location following their name by using a signature file.[15] They would also signal if they were currently traveling, attempting to account for their neglect of previous messages and inability to contribute in a fashion consistent with what they felt would otherwise be expected of them. These three techniques provided information that individuals thought would be useful for recipients in decoding the meaning they assigned to a message. They were further used in order to demonstrate interest in participation itself, a means for signaling engagement that would otherwise be difficult to detect in text.

Problems associated with terminology, relevance, and situational and background ambiguity accounted for the failure of the project to meet the expectations of its participants. However, these problems precipitated a set of adaptations to text-based interaction. Below we will argue that situations in which text is the primary medium for communication demand interpretative tools that account for the way language is used to organize interaction. Semiotics and, in particular, its applications in linguistics, offers a lens for understanding the above interaction problems *and* subsequent adjustments. We should first, however, consider alternative explanations for these problems.

Explanations Based on Interest, Commitment, and Incentives

This was a unique project. Panel members had committed to participation and signed a contract to interact online, making the expectations placed on them different from those in a usual face-to-face conference or academic e-mail discussion group. Yet the group composition and interaction formats in our study are commensurable with those in other research on group-based online interaction, allowing us to consider three common explanations for the problems our panel members encountered: the substantive, associative, and strategic arguments.

First, the substantive argument proposes that absence of topical interest might be responsible for the silence and other difficulties with the e-mail discussion. Rheingold (1993) claims that this factor determines issue-based interaction online. He argues that individuals will interact online when they share a common set of interests around a particular topic and are given the opportunity to discuss it.

The conversation topics pursued online were central to the theme of the project and the expertise of participants. Transcripts from the face-to-face meeting demonstrate their personal stake in discussing these subjects. The 12 panel members had dedicated their careers to activism and its analysis, achieving high levels of status within organizations considered to be at the apex of their professions. An analysis of participants' CVs demonstrates that each had dedicated an average of 15 years to research or advocacy around directly related issues. The offline meeting showcased their intimate understanding of the geographic areas with which they worked. Finally, 83% expressed a desire to engage in future projects on the panel's subject after the official end of the program. These facts make it unlikely that the communication problems can be attributed to disinterest in the issue at hand.

A second explanation might be called the associative argument. Perhaps the group was not committed to working together. According to Weber's (2004) arguments on open-source communities and Kollock's (1999) theories on the public provision of expert knowledge in e-mail discussion groups, online groups will face collective action problems similar to those described by Olsen (1965). These scholars assert that individuals have a disincentive to post to a group discussion, because the rewards from this action will be distributed among recipients. Silence, then, comes from skepticism about the ultimate payoff from participation.

This argument is challenged by our participants' attempts at interaction and the motivations they reported. E-mail transcripts show that individuals found the lack of interaction before and after the meeting problematic. Ninety-two percent of the participants indicated a "very high" or "high" desire to contribute to the e-mail discussion group. A surveyed participant reinforced this: "I had an interest in this network, as I was looking, among other things, for a sound form of continuity [of communication regarding information technology and activism] and I thought this group of people was an interesting starting point." Further, a set of posts that occurred immediately after the meeting indicates interest in continuing discussions initiated offline. And answers to the survey question "How important were the following factors in deciding to apply for inclusion on the project?" also contradict an associative explanation. The $2,000 honorarium and the status obtained from inclusion in the network were reported as extremely low motivations for application (although such professions of disinterest would be expected from high-status individuals). Instead, survey respondents claimed that they applied for admission in the hopes of discussing the topic of the panel, and 92% of participants answered that an "interest in engaging with other experts" was their dominant motive for applying for admission to the project.

Finally, reticence to interact might be seen as strategic, explained by a lack of professional incentive. Matzat (2004) argues that this is an important factor in online settings and finds that the desire to cultivate social contacts with other researchers motivates researchers' use of e-mail discussion groups. Consequently, group composition matters. This is similar to Bourdieu's (1984) argument that individuals in academic fields engage with those who offer the greatest professional returns for doing so. These scholars would

TABLE 1 Reasons for decisions to respond to e-mail discussion group posts

Factor	Amount of Influence*				
	1	2	3	4	5
Discipline of sender	. 75	17	0	0	8
Profession of sender	. 75	17	0	0	8
Interest in topic	. 25	8	8	25	33
Personal ties with sender	. 33	17	25	8	17
Time available	. 25	8	0	8	58

Note.—Data are percentages of respondents. Some rows do not add to 100 because of rounding.
* Influence scores range from 1, "not influential," to 5, "extremely influential."

expect people to be more likely to communicate with other researchers, as they have the most to gain through engagement or lose through inaction.

Table 1 presents reported motivations for e-mail discussion group behavior. Contrary to the expectations suggested by a strategic explanation, we see little connection between disciplinary or professional relationship and decision to respond to a post. If we subscribed to this explanation, we would also expect there to be a strong negative association between the decision to post and a desire to communicate across disciplines. Analysis of transcripts from e-mail discussion group and face-to-face interactions indicates that the sequences of responses among interactant dyads are not more likely to involve successive exchanges among individuals in the same discipline than among those in different disciplines.[16] And the low importance of personal ties for explaining motivation is not surprising, given the scarcity of preexisting relationships among group members.[17]

It is interesting to note that participants reported time constraints to be relatively influential on their decision to respond to a post. Considering this response as an indication of preferences in light of the opportunity cost of not engaging in other activities (Becker 1965), it suggests that group members felt that crafting an effective e-mail in this situation would require more work than they initially expected (see also Galegher and Kraut 1990). It is possible that this sentiment derives directly from the diverse disciplinary and professional affiliations in the group; the requirements of communicating across linguistic registers likely will produce more obstacles in e-mail than in person. The lack of immediate feedback and the inability to specify how one intends to be interpreted in e-mail may magnify difficulties and increase the time necessary to compose a message.

In this section, we have demonstrated that levels of interest in the subject, commitment to the project, and incentives for participation were all strong in the group we study. And yet these individuals were reluctant to invest the time and energy required to communicate effectively. Survey responses indicate that the demands of addressing the problems presented above could be too high online.

Notes

1. We compared the high- and low-volume posters according to whether they spoke English as a second language or not. Although not enough data are available for statistical analysis, there was no correlation between native language and propensity to post. (The three individuals who contributed most frequently spoke Italian, Spanish, and English, respectively, as a first language.)

2. The nonexperimental conditions of the study meant that we could not vary the group's composition around factors ordinarily studied by social psychologists, such as leadership, gender, task complexity, and status (e.g., Borgatta, Bales, and Couch 1954; Wheelan and Kaeser 1997; Ridgeway and Smith-Lovin 1999; York and Cornwell 2006). However, our group was less clearly task oriented than those investigated in the group interaction literature, as the participants were drawn together to informally discuss a set of issues important to their fields.

3. Intercoder reliability levels were high (Cohen's $\kappa = .85$) for most categories, and coding conflicts were mediated by a third researcher.

4. We attempted but were unable to reach several members because they did not respond to e-mails or had changed their address.

5. Participants were surveyed regarding their use of direct communication with others in the panel. Only three responded affirmatively, and they claimed they used direct communication for purposes unrelated to the project.

6. Questions were pretested through a cognitive interview approach (Winkielman, Knauper, and Schwartz 1991). Where relevant, response options were randomized to mitigate order effects, and bipolar scales were employed. Following Porter and Whitcomb's (2003) findings on the positive relationship between e-mail personalization and survey response rate, we contacted members individually by name and with messages reflecting individual characteristics. To reduce measurement error, the first possible answer in "drop box" questions was concealed (Couper et al. 2004), and graphics were used sparingly in order to minimize download time (Dillman et al. 1998).

7. In analyzing the data, we were conscious of the politics of transcribing complicated visual and vocal events to the printed page (Schieffelin and Doucet 1992; Goodwin 1994). We sought to record elements that captured our source of variation—i.e., the consequences of the use of different kinds of cues across contexts.

8. Members of the panel are highly literate users of e-mail: 75% subscribe to at least five other e-mail discussion groups, 67% contribute to these at least once a month, and 50% "provide assistance to the work of others" in e-mail discussion groups at least once a week.

9. A single arrow indicates that the following material is quoted from an e-mail message. Double arrows (‖) are frequently inserted by authors (or their e-mail programs) to indicate that they are quoting from a previous message. Typographical and grammatical errors remain in quoted e-mails in order to reproduce the group's conventions.

10. All names have been changed.

11. Emoticons are pseudolinguistic sequences of punctuation marks that depict an image of a face (usually smiling).

12. The smaller group that participated in this e-mail exchange was self-assembled among those who met at the offline meeting.

13. Group members maintained formality in the majority of their e-mails, beginning with greetings and ending with closures such as those used in conventional letters.

14. "Flaming" refers to "sudden, often extended flare-ups of anger" considered common to online interaction (Danet 1998). It has been reported both in laboratory settings and in a variety of business, governmental, educational, and public networks (e.g., Sproull and Kiesler 1986; Lea and Spears 1992; Thompsen and Ahn 1992).

15. A signature file contains text that follows the end of a message and, once created, is usually appended automatically to postings. The text may include contact information and institutional affiliation, a quote that is meaningful to the sender, or a legal disclaimer.

16. However, the interpretation of posting sequences is complicated by between-person differences in frequency of checking e-mail.

17. Note that the even distribution of the interest variable further weakens the plausibility of the substantive explanation discussed above.

Section 6

TMI

TMI

Too Much Information

By Lee Rainie and Barry Wellman

Perhaps the biggest complaint that people have in the era of networked information is that there is just too much information to monitor and digest. Still, the feeling of being overwhelmed by information did not start with the Internet and Mobile Revolutions. Historian Ann Blair has found scholars complaining as early as 1550 about a "confusing and harmful abundance of books."[25] However, what is unique is that ICTs provide not only more information but also more channels that connect people to this information and feed it to them. The unprecedented abundance of information that permeates the networked individual's life can often be difficult and stressful to manage. As one Connected Lives participant complained: "Searching [the] internet for information can be quite tedious, time consuming and not quite successful. You can spend hours trying to find one stupid fact and there is too much information, which is really hard to sort out."

To deal with TMI, networked individuals employ a number of strategies that range in complexity to cope with and manage the information overload. They rely on search engines, bookmarks, and tags. Moreover, people develop ways to alert them to new information about issues that matter to them. Pew Internet data show that two-fifths of Americans have set up news alerts through Google, Yahoo, news services, financial sites, and sports operations to update them every time a subject is mentioned on the web. Close to two-thirds get online newsletters related to work or hobbies. Some 37 percent have set up customized web pages to display information on subjects they care about –virtually all of which get up-to-date news of one kind or another about the people and the subjects that the creators have designated.[26]

Microblogging sites such as Twitter have led to yet another new way of managing information flows. With Twitter, networked individuals have the power to choose the people they want to follow and receive curated information from. Unlike Facebook, Twitter is asymmetric, so people can follow more (or fewer) people than follow them.[27] As entrepreneur Mark Suster blogged, "I follow really smart people from a wide variety of backgrounds and interests [and] they tell me what to read. … I pay attention to people I trust & respect and let them be my guides."[28]

Still, these strategies only begin to scratch the surface because they only involve gathering and disseminating information. Information assessment is another issue. With a deluge of information cascading from a variety of different sources, networked individuals must actively develop the skills to critically assess the institutional information they find and what they receive from their personal networks. The ability to balance these two information sources is a key for networked individuals as they cope with information overload.

Pew Internet conducted research in mid-2009 aimed at understanding the new patterns of engagement people had with their social networks and with media in a particularly important context. Through a national telephone survey, online interviews, and in-depth phone interviews, Pew Internet researchers asked people about the ways they were getting information and advice about the 2008 recession and its lingering effects.[29] There was a clear sense in the survey data that people were trying to make sense of complex economic problems that were not easily explained by traditional economic theories. As such, Pew Internet asked about five specific sources of information and support that may help networked individuals to better understand the general economy and their own personal finances (table 9.1). For information about the general economy, a great majority of Americans were most likely to seek out traditional news sources in broadcast and print media (84 percent and 64 percent, respectively). When it came to the two-thirds (64 percent) of the sample who had home broadband internet connection at the time, the internet became more prominent as an information source about both the general economy and one's personal finances. Significantly, use of the internet did not displace people's reliance on interpersonal networks of friends and family.

The survey showed that during this time of uncertainty, networked individuals balanced both institutional and interpersonal information. People did not either talk to

Table 9.1 Sources used for information about the economy and personal finance in the United States

Sources of Information	General Population		Those with Broadband at Home (64% of Sample)	
	General Economy	Personal Finance	General Economy	Personal Finance
Television and radio	84%	45%	85%	46%
Newspapers, magazines, books	64	44	67	43
Internet	48	38	67	52
Friends and family	40	37	45	40
Financial professional	17	24	21	28
None of these sources	6	20	5	18

Source: Pew Internet & American Life Project.

others or consult a single media platform. Rather, they foraged among sources and communicated with a range of people. As the recession took hold, the average American used two or three sources of information to make sense of what was happening and to plan personal coping strategies. People talked to other people, sought updates from media sources like newspapers and broadcast media, and actively searched for insights that would help them explain what was happening to the economy and how they might adjust to those changes.

The behavior of networked individuals permeates the survey results. One typical example is Sharon Hockensmith, the sixty-six-year-old wife of a retired Air Force officer. Sharon was attuned to market vibes in part because of her nature as an information omnivore. During the time of the recession, she and her husband signed up to receive email newsletters and online alerts from several financial companies. They also subscribed to some financial blogs with a free market emphasis. Moreover, they began to watch cable TV financial shows like *Fast Money* and *The Kudlow Report.* In short, Sharon juggled networked information the way a networked individual would: "We research companies online; check with our investment adviser; exchange ideas with a couple of financially-savvy friends, and when we know enough, we make investments that, more than ever before, take into account the way the political wind is blowing."

Another example of a networked individual balancing between institutional and interpersonal information sources was a Pew Internet respondent who was looking to buy a new home:

> We had been in the market to buy a home for a little less than a year. As the housing bubble burst our neighborhood (Park Slope, Brooklyn) saw a stagnation in housing prices. We felt with our rental lease up we should pounce. We used NYTimes.com to check out open houses. That came after a year in which we had talked to a lot of brokers to get a sense of what we needed to ask. We used online forums (Brooklynian.com and others) to find out about the area and get recommendations for our lawyer. We talked with our parents and siblings (home owners) and we talked with friends working in finance to determine if it really did make sense to buy. After all this work, we decided it did. We found a place being sold by someone not using a broker. We got approved for the mortgage and got the place.

Thus, networked individuals use a number of strategies to help manage the abundance of information that is available to them both online and offline and they exploit both institutional and interpersonal information to help them in their everyday decisions.

The "Veillance" of Personal Information

Beyond general information and news, however, much of the content that is being unleashed in the digital world is that of networked individuals' personal information.

The increasing popularity of social networking sites, especially Facebook, has resulted in the public sharing of sensitive personal information such as one's location, marital status, workplace, contact information, and many other details. Users are at least partially aware of some of the information about them that is available online. In a national survey in September 2009, Pew Internet found that 57 percent of internet users had used search engines or other search strategies to see if there was material about them online and 63 percent of them had found at least something about themselves.[30] Among all internet users:

- 42 percent know a picture of them is available online
- 33 percent know their birth date is listed online
- 31 percent know their email address is listed online
- 26 percent know their home address is listed online
- 23 percent know that something they have written is listed online
- 22 percent know the groups or organizations they belong to are listed online
- 21 percent know their home phone number is listed online
- 12 percent know their political affiliation is listed online
- 10 percent know a video of them is available online
- 44 percent of employed internet users know the name of their employer is listed online
- 12 percent of the internet users who have cell phones know their cell number is listed online

When shared online, personal minutiae can be offered by networked individuals to build trust and make online interactions more efficient. Though this brings benefits, it also impacts people's privacy. "Surveillance" is a commonly used word, adapted from the French. But the rise of networked information in the Internet Revolution has enhanced two other forms of "veillance," or "monitoring": peer-to-peer "coveillance" and "sousveillance" by ordinary people of those in authority.

Surveillance

The social life of digital information has opened up the doors to new means of surveillance by government and organizations. Through monitoring social media, governments have found a new way to keep an eye on the behaviors and actions of their citizens. In China, for instance, the Ministry of Public Security has developed an extensive and sophisticated system of surveillance that limits access to information that Chinese leaders believe may disrupt the state's stability or undermine security. In the words of Greg Walton, a researcher for the International Centre for Human Rights and Democratic Development: "Old style censorship is being replaced with a massive, ubiquitous architecture of surveillance ... [and] the aim is to integrate a gigantic online database with an all-encompassing surveillance network—incorporating speech and face recognition, closed-circuit television, smart cards, credit records, and internet surveillance technologies."[31]

To take one important example, Chinese regulations require all internet service providers to record for at least sixty days the identities of their users, the websites they access, the time they spend on those websites, and any other online activities.[32] That information is handed over to government officials when requested. More than monitoring the websites accessed, the government also scrutinizes the electronic communications of its citizens. Thus the government works with TOM-Skype, the Chinese version of Skype, to gather users' private voice, video, and text conversations. They regularly scan chat messages for specific keywords that are deemed offensive or politically sensitive.[33]

China is far from being the only state exercising such surveillance practices. Western democracies, including the United States, also take part in such activities. For example, following the September 11, 2001 terror attacks that gave rise to Americans' safety concerns, the US government instituted an eavesdropping program to collect both domestic and international communications. Corporations, too, are taking advantage of technological advances to gather information about consumer habits, behaviors, and interests as a means of turning a profit.[34] Writer Evgeny Morozov quips: "The only difference between the two is that one system learns everything about us to show us more relevant advertisements, while the other one learns everything about us to ban us from accessing relevant pages."[35]

Ironically, corporations are exploiting the very same systems of aggregation of user data that were noted earlier as creating new signposts of credible and trustworthy sources. They are using internet-tracking technologies to collect information about networked individuals' online activities, behaviors, attitudes, buying habits, and interactions. For instance, the *Wall Street Journal* found that America's top fifty websites installed an average of sixty-four pieces of tracking technology onto computers resulting in a total of 3,180 tracking files.[36] These data are often commodified and sold to the highest bidder to help businesses market services and products. One telling example discovered by the *Journal*: A tracking "cookie" surreptitiously installed by Lotame Solutions on Ashley Hayes-Beaty's computer that consists of a single code—4c812db292272995e 5416a323e79bd37—accurately identified her as a twenty-six-year-old female in Nashville who has searched for information about the movies *The Princess Bride* and *50 First Dates*. The cookie knows that she has watched the *Sex and the City* TV series, browses entertainment news, and likes to take quizzes.[37]

Hackers or criminals can also gather personal information such as location tagging or status updates about one's daily activities. To demonstrate this, Dave Marcus, director of security research and communications at McAfee Labs, has followed people through the location tagging of their tweets and documenting their schedules, place of work and home—many of which were unwittingly provided online by the individuals themselves.[38] The website PleaseRobMe.com, established in 2010, also brought attention to the dangers of providing such information online. Aggregating and live-streaming publicly shared check-ins via foursquare and Twitter, PleaseRobMe.com showed when people left their homes: pointing out just how easy it is for technologically savvy and determined criminals to accomplish their goals.[39]

Coveillance: We Watch Each Other

Ordinary citizens now frequently engage in practices of "coveillance," which people use so they can observe each other.[40] Search engines and social networking sites are the primary sources people use to find out more about both known and unknown individuals.[41] While they may be unhappy about others checking up on them, Americans are quite willing to exploit the internet to check up on others. In a 2009 Pew Internet survey, 69 percent of internet users reported searching for someone online, up from 30 percent in 2001 and 53 percent in 2006 when similar surveys were conducted. The later survey found that:

- 46 percent of internet users had searched for someone from their past or someone they had lost touch with
- 44 percent had searched for someone whose services or advice they were seeking in a professional capacity like a doctor, lawyer, or plumber
- 38 percent had searched for friends
- 30 percent had searched for family members
- 26 percent had searched for coworkers, professional colleagues, or business competitors
- 19 percent had searched for neighbors or people in their community
- 19 percent had searched about someone they just met or someone they were about to meet for the first time
- 16 percent had searched for people they were dating

What do they search for? Contact information (69 percent), social network site profile information (48 percent), photos (43 percent), information about professional or career accomplishments (36 percent), personal background information (27 percent), public records related to things like real estate transactions, divorce proceedings or bankruptcies (27 percent), and whether someone is single or in a relationship (17 percent).

Tracking others in this manner can seem creepy. Indeed, the terms "Facebook stalking" and "creeping" have been coined to describe the act of using Facebook to find out more information about those within or even outside of one's personal network.[42] As one student says of this endemic practice: "There's only so much you can learn when you first meet someone. By Facebook stalking, I can learn more about the person, like, who they're dating, their interests, any common friends that we may have and random tidbits of information you wouldn't get on your first encounter."

Scholarly research agrees: Facebook stalking has become so prevalent in the lives of teens and young adults that a Facebook page called "Facebook Stalking. ... Admit it, you do it" has more than 820,200 "likes" in August 2011.[43] More than that, a Google search of the keywords "Facebook stalking tips" pulls up a number of blogs and articles of do's and don'ts, anecdotes, and best practices. The increasing popularity of online dating websites has also resulted in more coveillance as individuals looking for romantic partners must assess the credibility and reduce the inevitable uncertainties of encountering others online. In 2010, nearly one-quarter (23%) of American online daters engaged in information-based triangulation: checking public records and cross-referencing and comparing profiles on multiple websites.[44]

There is coveillance in other realms of life. In interviews tied to its 2009 survey on reputation management, Pew Internet heard from a large number of those who learned important things in their searches about others. One woman who was adopted as an infant wrote:

> I used the internet to trace my birth father through a search. He gave me the name of my birth mother. Through a combination of in-person research and online queries, I patched together the history of my birth mother through property records, birth records, divorce records, and genealogical records (especially a family history placed online by a birth-great uncle). [T]hrough all of this, I found my birth mother, who refused contact.
>
> A few years later I found my birth sister. She is now one of my best friends. We look alike and are alike in many ways. It turns out she vaguely knew about me and looked on the internet for me and came close. We both are very happy to have met each other and it would probably have never happened without the internet.[45]

Other respondents to the Pew Internet survey told of tracking down damaging personal information on a pastoral candidate who was being recruited by their church; described how a search about a physician giving important, intimate advice provided details that he is a transsexual; revealed that a man who used to date a Pew respondent's sister was an avid participant in "furry fandom" events where people dress as animals that exhibit human personality traits; that a dentist who had been wrongfully billing the Pew respondent had also been overbilling other clients; that a boss had quietly accepted a job at a competing firm; and that a would-be tenant was a convicted pedophile. Many described tracking down old flames and long-lost friends. Some described learning too much information about the sexual adventures of younger relatives.

Some of the most riveting stories came from Karyl Chastain Beal about the website she runs for those who want to memorialize suicide victims, suicidegrief.com.

> A woman named Melissa submitted her own name for our website's memorial wall. I only knew she lived in Illinois. I searched and found where she lived and I contacted the police. They got to her house before she died. She was getting ready to take an overdose of pills. … In another similar case, a woman sent a note to one of the suicide-watch groups I run. She sent a

suicide note through the group to me. I used Google to track her down in Canada and called the Royal Canadian Mounted Police. They found her on the floor; she had already taken the pills, but fortunately they got her to the hospital in time.

One of the consequences of all this self-monitoring and tracking of others online is that it increases people's awareness of very weak ties—and that likely changes the way they mobilize our networks. It gives people a better sense of the potential power of their networks and the specific people who might help them address a problem—whether that problem is finding a cancer specialist or a new job. The disclosures and revelations that would have previously been shared with only a small, intimate network of family and friends (or not at all) become valuable indicators of the professional and personal competencies among networked individuals. Such enhanced awareness also gives networked individuals more information than they might otherwise have about such things as the political views, the cultural tastes, the friendship circles, the basic lifestyle preferences, and even the daily activities of those in their networks.

Sousveillance: Watching the More Powerful

In direct opposition to panopticon surveillance where organizations observe people from on high, "sousveillance" is the observation from below of more powerful organizations and people. Steve Mann invented the term when he decided to watch the watchers by video blogging his interactions in department and chain stores, including the surveillance cameras on their ceilings.[46] But most of the sousveillance action is now on the internet, where networked individuals can now find information that has the potential to destabilize power relations.

Wikileaks.org has undertaken the most controversial and publicized sousveillance. It is an organization that releases confidential governmental information online from anonymous news sources who submit sensitive material to an electronic drop box. Its motto is: "We help you safely get the truth out." In October 2010, the organization leaked approximately four hundred thousand private and classified documents that became known as the Iraq War Logs. Following that, in November 2010, Wikileaks began to release and publish a total of 251,287 secret diplomatic cables from US embassies around the world dating from 1966 to February 2010. The release of the cables revealed controversial foreign strategies, such as a secret intelligence campaign where information such as passwords, credit card numbers, email addresses, and even biographic and biometric data of United Nations leaders were collected.[47] Although proclaiming its openness to multiple contributors, the controversial nature of the site—and the probability of government surveillance of its contributors—has reduced contributions. Indeed, the Wikileaks site would not even open on April 9, 2011, and the Wikipedia's "Wikileaks" article reports that "the wikileaks.org domain redirected to mirror.wikileaks.info."[48]

Surveillers can also be sousveilled, as when the OpenNet Initiative works to "investigate, expose and analyze internet filtering and surveillance practices" of seventy countries.[49] It provides information about their filtering and censorship practices as well as the legal, technical, and administrative tools they use.[50] Its sister project, the Information Warfare Monitor, uncovered the GhostNet cyber-espionage network emanating from within China that targeted the computers of the Tibetan community in 2009, including the private office of the Dalai Lama.[51]

The different manifestations and levels of "veillance" show that networked information, more specifically personal information, is bound to privacy concerns. A variety of actors can now more easily exploit the wealth of information available online in ways that fulfill their respective needs. For governments, this means watching over citizens to ensure the "stability" and "security" of the state, while for businesses, this means collecting data about consumer behaviors to find new ways of making a profit. Ordinary citizens may also watch each other side by side in an attempt to find more nuanced information about family, friends, acquaintances, employees and employers, prospective romantic relationships, and even strangers. And of course, sometimes the purpose may be to challenge the surveillance practices of authorities, essentially "surveilling the surveillers." In this way, networked information lives a social life that is deeply complicated and heavily layered.

Dealing with the "Zero Privacy"

The ever-thorny and increasingly salient issue of privacy has been brought to the forefront of popular discourse as networked individuals share information about themselves and as governments, organizations, businesses, and individuals have more power to watch over one another. Former CEO of Sun Microsystems Scott McNealy famously said, "You have zero privacy ... get over it."[52] Facebook founder and CEO Mark Zuckerberg claims that "People have really gotten comfortable not only sharing more information and different kinds, but more openly and with more people. That social norm is just something that has evolved over time."[53] His company followed that belief by giving third-party application developers access to each member's data.[54]

As surveillance, coveillance, and sousveillance proliferate, people have become more aware of the issue of privacy. The evidence shows that networked individuals, both adults and youth alike, would like to keep the norm of privacy alive by controlling the information going out both to their networks and the wider public. In May 2011, a Pew Internet study found that 58 percent of the adult users of social networking sites set their accounts so that only friends can see what they post and another 19 percent use settings to make their account partially private. A quarter of those who restrict access to only their friends (26 percent) have taken the further step of limiting what their friends can see. Another strategy people use to control their identity is to use different profiles on multiple social networking sites: 42 percent of social networking site users have profiles on at least two sites and another 8 percent have more than one profile on the single site they use. Moreover, despite concerns that sites such as Facebook make it hard for users to adjust their privacy controls, 79 percent of social networking site users said they found the privacy-setting systems not difficult or not too difficult to use.[56] These strategies show that adult internet users are trying to manage their identity to some extent.

danah boyd and Eszter Hargittai found that in response to these issues of surveillance, young people are actively seeking ways to protect themselves and control the information they release to the public. For example, about one-quarter (24 percent) of all Facebook users changed their privacy settings four or more times in 2009, with this number increasing to more than half (51 percent) in 2010.[57] Similarly, Pew Internet surveys have found that 66 percent of all teens with an online social networking profile have restricted access by making the profiles private, adding password protection, hiding them entirely from others, or even taking them offline.[58] Moreover, tech-savvy youth are creating specialized sublists of friends to further manage information flows.[59] One student explained, "I put a lot of my family members, especially the older ones like aunts and uncles, on my Limited Profile list so that they don't see my pics or my status updates. Sometimes I have pics at parties and swear on my updates and they just don't need to see that."

Other social media users are even removing friends with whom they no longer want or feel obliged to stay in touch.[60] As Pew researchers Amanda Lenhart and Mary Madden conclude: "For teens, all personal information is not created equal." They filter the personal information that they share with particular others, controlling who is able to see what on the basis of the nature of the tie and the particular circumstances.

The evidence for both adults and youth show that there is at least some deliberate attempt at controlling what personal information is released on the internet and for whom specifically. Networked individuals are aware of the costs that come with giving unfettered access to their personal information online and thus adjust their online behavior accordingly. Of course, there are still the covert methods of surveillance practiced by governments and organizations. But in the face of these mounting challenges against privacy, networked individuals are trying to find ways of adapting to surveillance and coveillance.

Notes

25. Ann Blair, "Reading Strategies for Coping with Information Overload ca. 1550–1700," *Journal of the History of Ideas* 64, no. 1 (2003): 11–28, quotation on p. 11.

26. Kayahara and Wellman, "Searching for Culture," see note 24; see also David Weinberger, *Too Big to Know* (New York: Basic Books, 2012).

27. Fallows, "Internet Searchers," see note 15.

28. Mark Suster, *Both Sides of the Table*, December 20, 2010, http://www.bothsidesofthetable.com/2010/12/20/the-power-of-twitter-in-information-discovery.

29. Lee Rainie and Aaron Smith, "The Internet and the Recession," July 2009, http://www.pewinternet.org/Reports/2009/11-The-Internet-and-the-Recession/3-How-the-internet-and-other-sources-have-helped-people-cope-with-the-recession/1-Americans-have-used-several-sources-of-information-and-advice-in-the-recession.aspx.

30. Mary Madden and Aaron Smith, "Reputation Management and Social Media," Pew Internet & American Life Project, May 26, 2010, http://www.pewinternet.org/Reports/2010/Reputation-Management.aspx.

31. Greg Walton, "China's Golden Shield: Corporations and the Development of Surveillance Technology in the People's Republic of China," http://www.dd-rd.ca/site/_PDF/publications/globalization/CGS_ENG.PDF.

32. Article 14 of the government of China's Measures for Managing Internet Information Services, http://www.chinaculture.org/library/2008-02/06/content_23369.htm.

33. Nart Villeneuve, "Breaching Trust: An analysis of surveillance and security practices on China's TOM-Skype platform," http://www.scribd.com/doc/13712715/Breaching-Trust-An-analysis-of-surveillance-and-security-practices-on-Chinas-TOM Skype-platform.

34. The *Wall Street Journal*'s "What They Know" series reports on the Internet-tracking technologies being used by businesses, http://online.wsj.com/public/page/what-they-know-digital-privacy.html.

35. Evgeny Morozov, *The Net Delusion: The Dark Side of Internet Freedom* (New York: Public Affairs, 2011), p. 97.

36. *Wall Street Journal*, "What They Know," see note 34.

37. Julia Angwin, "The Web's New Gold Mine: Your Secrets," *Wall Street Journal*, July 30, 2010, http://online.wsj.com/article/SB10001424052748703940904575395073512989404.html.

38. Lynn Greiner, "The Perils of Social Networking," *ComputerWorld Canada*, October 18, 2010, http://www.itworldcanada.com/news/the-perils-of-social-networking/141749.

39. PleaseRobMe.com has stopped showing these check-ins after making the point that "if you don't want your information to show up everywhere, don't over-share."

40. The term "coveillance" was invented by Barry Wellman. See Steve Mann, Jason Nolan, and Barry Wellman, "Sousveillance," *Surveillance and Society* 1, no. 3 (2003): 331–355.

41. David Westerman, Brandon Van Der Heide, Katherine Klein, and Joseph Walther, "How Do People Really Seek Information about Others?" *Journal of Computer Mediated Communication* 13 (2008), no. 3: 751–767.

42. Bonnie Ruberg, "10 Signs You've Officially Become a Facebook Stalker," January 5, 2009, http://www.heartlessdoll.com/2009/01/10_signs_youve_.officially_become_a_facebook_stalke.php.

43. Facebook.com, "Facebook Stalking," http://www.facebook.com/pages/Facebook-Stalking-Admit-it-you-do-it/147838687575.

44. Madden and Smith, "Reputation Management and Social Media," see note 30.

45. Jennifer Gibbs, Nicole Ellison, and Chih-Hui Lai, "First Comes Love, Then Comes Google," *Communication Research* 38, no. 1 (December 2010): 70–100.

46. The term "sousveillance" was invented by Steve Mann. See Mann, Nolan, and Wellman, "Sousveillance," see note 40.

47. Wikileaks Secret US Embassy Cables database available at http://wikileaks.org/cablegate.html.

48. Robert Booth and Julian Borger, "US Diplomats Spied on UN Leadership," *The Guardian,* November 28, 2010, http://www.guardian.co.uk/world/2010/nov/28/us-embassy-cables-spying-un; *Wikipedia,* "Wikileaks," 2011, http://en.wikipedia.org/wiki/Wikileaks.

49. The National Institute on Money in State Politics was founded in 1999 and "dedicated to accurate, comprehensive and unbiased documentation and research on campaign finance at the state level," http://www.followthemoney.org; Scott Walker information from http://www.followthemoney.org/database/StateGlance/candidate.phtml?c=116585.

50. The OpenNet Initiative is a collaborative project between the Citizen Lab at the University of Toronto, the Berkman Center for Internet and Society at Harvard University, and the SecDev Group: http://opennet.net.

51. OpenNet Initiative Regional Overviews are available at http://opennet.net/ research/ regions; Information Warfare Monitory, *Tracking GhostNet: Investigating a Cyber Espionage Network,* March 29, 2009, http://www.scribd.com/doc/13731776/ Tracking-GhostNet-Investigating-a-Cyber-Espionage-Network.

52. Sun Microsystems' former CEO Scott McNealy said this as early as 1999. See http://www.wired.com/politics/law/news/1999/01/17538.

53. Zuckerberg, Remarks at the *Crunchie Awards,* January 2010, http://www.computerworldxom/s/article/9143859/Facebook_CEO_Zuckerberg_causes_stir_over_privacy; see also http://crunchies2009.techcrunch.com/about.

54. David Kirkpatrick, *The Facebook Effect* (New York: Simon & Schuster, 2010).

55. Interview with Eric Schmidt, "Google's CEO: "The Laws Are Written By Lobbyists," *The Atlantic,* October 1, 2010, http://www.theatlantic.com/technology/archive/2010/10/ googles-ceo-the-laws-are-written-by-lobbyists/63908.

56. Mary Madden and Aaron Smith, "Reputation Management Online," September 2011.

57. danah boyd and Eszter Hargittai, "Facebook Privacy Setting: Who Cares?" *First Monday* 15, no. 8 (2010): http://firstmonday.org/htbin/cgiwrap/bin/ojs/index.php/fm/article/view/3086/2589.

58. Amanda Lenhart and Mary Madden, "Teens, Privacy, and Online Social Networks," April 2007, http://www.pewInternet.org/Reports/2007/Teens-Privacy-and-Online-Social-Networks/1-Summary-of-Findings.aspx.

59. Brady Robarts, "Randoms in My Bedroom: Negotiating Privacy and Unsolicited Contact on Social Network Sites," *PRism 7,* no. 3 (2010), http://www.prismjournal.org/fileadmin/Social_media/ Robards.pdf.

60. See note 59.

Cell Phones and Email

By Christena Nippert-Eng

Today's explosion of information and communication technologies (ICTs) and their widespread adoption by so many people have exponentially complicated this reality. Kenneth Gergen (1991) has offered what is perhaps the most seminal work on this to date. Today, he argues—largely because of communication technologies—many people engage in far more relationships than the average individual would have had at any previous point in history.

In fact, Michael Schrage (1997, 1) perfectly captures the importance of these technologies for relationships. When it comes to information technology, Schrage tells us, "Whenever you see the word 'information' … substitute the word 'relationship'" to more fully understand its uses and its consequences. Claude Fischer (1992, 268) concurs in his excellent social history of the telephone, arguing these are indeed "technologies of sociability." But, he notes, expanded sociability is not all positive: "a key draw back of the home telephone is that very same expanded sociability. To have access to others means that they have access to you, like it or not."

The number and kinds of demands for attention that we are likely to receive at any given time, in any given place, are much greater when these technologies are in use than when they are not. Mobile communication technologies not only facilitate this burgeoning request for attention, they add another twist. They let others outside our immediate physical grasp reach us, but they also let them do so in what have been traditionally and especially interstitial places and times—where and when, for most of history, it would have been very difficult if not impossible for this to happen.[12]

The implications of this for the sheer amount of privacy and relationship work that we must do are staggering. Simmel (1955) was the first to point out that the modern sense of self is defined by memberships in a unique and diverse "web of group affiliations." One implication of this is that the modern individual's social territory is likely to include a larger number of more diverse social relationships than those that are possessed by members of premodern social groups.

In fact, Gergen (1991) argues that today—in what some call the post-modern society—the relationship landscape has virtually exploded. It is not simply that we have more relationships demanding even more attention because of our technologies, either. Rather, our contacts with others are scattered across far more channels of (often instantaneous) communication than ever before, too.

First, these technologies are well suited to blurring any subterritories we may have created between and within our social worlds, however unintentionally. The result is a breakdown of the traditional segmentation of roles, social networks, and personal ways of being that many people have long relied on in the U.S. to help regulate their accessibility—and their attention.[23]

Gergen (1991) argues that one of the effects of these technologies is that we become aware of the multiplicity of selves, of the many identities of which we are capable. This presents a bit of a problem when communications with others evoke or demand different identities and ways of being than those that we are actively engaged in at the moment. Depending on the nature of a contact, we may find ourselves required to instantly transform our current frame of mind in order to accommodate whatever mentality is mandated by a newly appearing request.[24]

The effect of a communique that instantly appears in a different place can be quite jarring at first. This was certainly the case for the segmenting individuals I studied whose "work" and "home" selves were distinctly embedded within specific times and spaces. Meyrowitz (1985) found a similar effect, too, in his study of the distinct identities, social groups, and interactions associated with specific times and places prior to the advent of television. In both cases, though, we found that cross-realm communications gradually chip away at the social-psychological walls separating social worlds; more integrated lives and identities result.

Thanks to our use of communication technologies, many more individuals may be experiencing both of these effects—the discomforting, cross-realm conflicting one and the boundary-challenging, possibly identity-merging one. People associated with one aspect of our lives are now suddenly requesting our attention with increasing regularity in all kinds of places and times other than those in which they traditionally might have been expected to appear. This is not limited to cases of "cellular interruptus," either. The result is that numerous boundaries are far less certain today and rules about one's accessibility across them are no longer so obvious—or easily assumed.[25] All it takes is a few people to selectively ignore one's wishes and we quickly discover that mobile technologies especially can make managing our accessibility a very personal burden, requiring a more personal solution than ever before.[26]

This is happening in part because each individual seeking our attention has her or his own ideas of where and how these lines should be drawn. This includes individualized preferences about not only how much access and attention we should give them, but of the forms that our accessibility and attention should take, too. Informing each of these individualized preferences are marked changes in related practical expecta- tions about things like the sheer number of responses we should be prepared to make (or demand), the speed of those responses (our own and others'), and the plethora of preferred response modes that we are all supposed to manage.

What emerges is a picture of what can be a daily, extremely personal, and constant juggling act. It's a never-ending and dynamically situated problem of whom and what to pay attention to, when, and in what way. The scope of the problem encompasses our face-to-face and physical engagements, as it has for quite some time. But now it also includes those that occur in the ever-larger, increasingly compelling, more virtual spaces of our lives.

Of course, as Rob Shields (2003) points out, no virtual encounter happens without it simultaneously being embedded in a physical one. In fact, as Gergen notes especially in his essay "The Challenge of Absent Presence" (2002), the demands and opportuni- ties for our attention that appear via communication technologies compete directly with and often supplant the giving of one's attention to things happening within one's physical presence. Indeed, Chayko (2008, 181) argues that "the younger among us may not even experience a sharp distinction between online and offline phenomena, and it would not be unreasonable to speculate that in time the online-offline distinction may fade entirely." Certainly, the more the playing field is defined as including everyone who is mentally present—whether they are physically present as well or must be ac- cessed using the technological device in one's hand—the more complicated the whole attention-demanding balancing act gets. Thus, the ongoing negotiation of who matters most and least in any given moment demands ever more of our constant attention.[17]

Table 3.1, "Number of self-reported communication technology access points for each participant," presents information on the number and kinds of channels partici- pants used at the time of this study. It shows how individuals today need to navigate between a number of telephones with different locations and numbers, email accounts, beepers, pagers, instant messages, and text messages. The average person in our subset had five technology access points through which they could be reached any given day—three telephone numbers and two email accounts.[18]

An individual's multiple access points—and the channels to which they are at- tached—become stitched together to create worlds of persistent access. The effect of this on daily life can be daunting, if not overwhelming. As Gergen (1991, 75) puts it, "In effect, the potential for new connection and new opportunities is practically unlimited. Daily life has become a sea of drowning demands, and there is no shore in sight." Or, as one participant put it,

> I'm a slave to my cell phone because people can reach me anywhere, anytime.
> I started hearing my cell phone ringing when it's not ringing. It's really freak- ing me out. I really want to pitch it.

This is the overall context of the remainder of this chapter, then, where, for at least some people, daily life—and the problem of achieving privacy—includes (1) a

TABLE 3.1 Number of self-reported communication technologies access points for each participant

Case #	Total Phone Numbers	Home	Work	Mobile	Other	Beeper	PDA	Total Email Addresses	Personal	Work	Other	Other	Total Access Points
30	5	1	2	2				3					8
40	2		1	1				2	1	1			4
41	2	1		1				2					4
42	2	1		1				4	2	1	1		6
43	2	1		1				1					3
45	5	1	3	1				2	1	1			7
46	2	1	1					2					4
47	5	1	3	1				1					6
48	2	1		1				1					3
49	3	1	1	1				1					4
50	2	1	1			1		1		1			4
51	3	1	1	1				2					5
52	5	2	2	1				2					7
53	3	1	1	1				1					4
54	3	1	1	1				1					4
55	2	1	1					1					3
56	3	1	1	1		1		3	1	2			7
57	2	1	1					4					6
58	4	1	1	1	1			1					5
59	2	1		1				1					3
61	2	1	1					2					4
62	3	1	1	1				2					5
65	3	1	1	1		1		2					6
66	3	1	1	1				1				1	4
67	2	1		1				1					3
68	3	1	1	1				2					5
69	2	1		1				1					3
70	2	1		1									2
81	3	1	1	1		1		12					16
84	3	1	1	1									3
85	2	1		1				2					4
89	2							1					3
92	2	1		1		1		1					4
Total	33	91	32	28	28	1	5	8	63	5	6	1	1	159
Averages		3	1	1	1	0	0	0	2	0	0	0	0	5

burgeoning number of demands for their attention, appearing across an increasingly diverse communication landscape; (2) a marked increase in the number of relationships they have or might have with the people making those demands; and (3) a shrinking set of shared rules, expectations, and practices for how to manage all of this.

In addressing this reality, the issue of power and its agenda-setting dimension comes to the fore: how much weight do we give to our own ideas of what we want to do—and when and how we wish to do it—versus other people's ideas of what we ought to be doing? Clearly, this begs the question of how difficult it might be for an individual to figure out what, exactly, she or he should pay attention to at a given moment. Should it be the baby one is bathing or the person on the other end of the ringing telephone? The individual on the office phone or the one on your cell phone? The person who wrote you the first email of the day, the author of the most recent one, or the one whose demand is most easily answered? Should it be any of the two hundred unread messages you've just seen on the computer in your hotel's lobby while on vacation or should it be the family waiting for you at the pool? The report that needs writing or the constant stream of IM questions that keep your subordinates working? Your pager or the patient waiting in the examination room? The list of options is seemingly endless.

The actual process of assigning priority to the very different demands for one's attention is rather mindboggling—and very likely to get any of these participants into at least a little trouble no matter what choice they make. It is also well outside of the scope of this book. Rather, here I limit myself to acknowledging the difficulty of the process and pointing out that this is precisely why developing techniques, strategies, and practices to manage these demands is so important.

Channeling: Prereceipt Management of Demands

Before a specific demand even reaches them, participants try to sort these requests into various channels of communication. Through more active and passive ways, they direct people to use different lines of communication to reach them. They follow more general as well as situational guidelines as they allocate people to these different channels.

Consider this participant, for instance, a consultant who frequently but not always works from home. He describes the specific ways in which he allocates people across his system of email, IM, a home phone, and a cell phone, granting access in specific ways and denying it in others in order to better regulate his attention. It's a good example of how participants draw on the full breadth of their communication systems to better control the demands others make on them.

> I have to say I'm fairly new to using it [instant messaging]. I literally had never used it until eight months ago. Um. And the group of people I use it with is still very small. It's like a total of maybe ten. So we all, I think, use it in a very similar way.

SENDER	PRIMARY & ADITIONAL FILTERS		HOLDING PLACE	RECIPIENT RESPONSE RATE
	filter #1 (channel)	filter #2, etc. (screener)		
Child's Caregiver	CELL PHONE	RING	CELL PHONE VOICE MAIL	fastest
Coworker	WORK PC	SPAM FILTER	INBOX	
Business Contact	WORK PHONE	SECRETARY	PERSONAL MAILBOX (MAILROOM)	
House Painter	HOME PHONE	ANSWERING MACHINE	FAMILY ANSWERING MACHINE + (e.g.) WIFE'S HEAD	
Online Retailer	HOME PC	SPAM FILTER	FAMILY EMAIL ACCOUNTS + (e.g.) HUSBAND'S HEAD	slowest

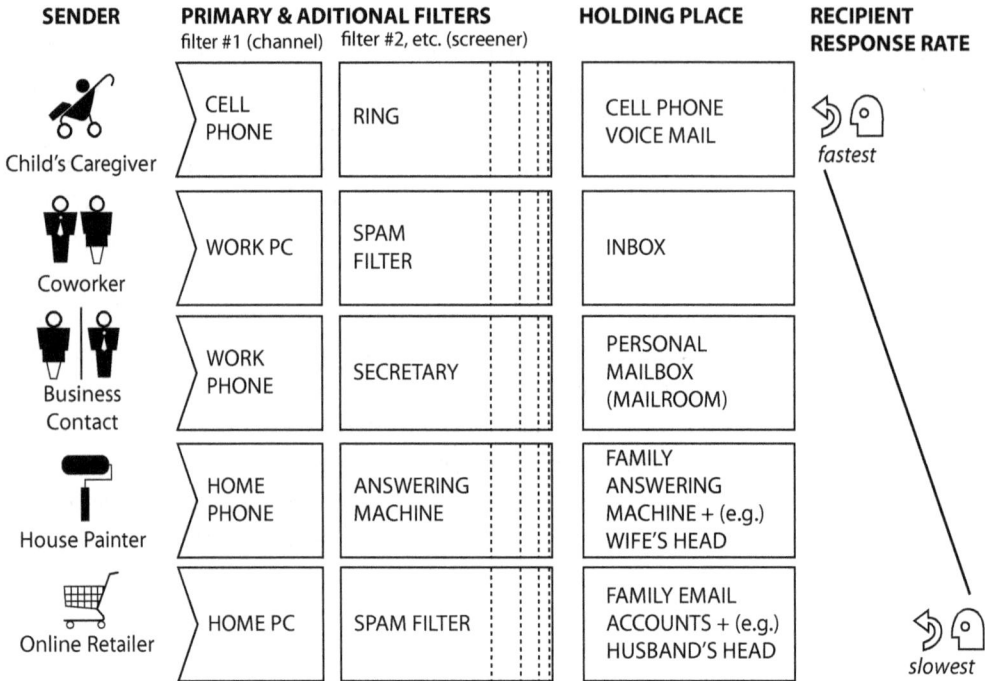

Figure 3.1 Channeling and the pre-receipt management of ICT-delivered demands for attention.

So those are other consultants?

Yeah. Or coworkers, people I am working on jobs with, or whatever, but—

So how do you find out that they have instant messaging?

Well, you just tell them, It's like, "This is my instant messaging address, add me to your list."

… So do you give somebody your email address and your instant messaging address at the same time?

No. I only give out that [IM] if we have some sort of established relationship. And as we go. I think, "Oh, if this person IMs me it would be good," I don't usually right off say, "Hi, I have this IM address and that's how you should get to know me."

Do people do that to you? Do people—

I think I have seen that a couple of times, maybe. But, it's the whole, like, what do you put on your business card? A lot of people want to put everything. They put their home phone, their babysitter's pager, whatever. That's exaggerating. But, um—And their IM and all that.

But, um, I like to—I used to not even want to put my phone on my business card. It kind of depends on your job function. I would just write that in if I needed it.

So I treat IM much the same. I'll add it if I think I need it, but by default, email is the only thing I give out. It's like—discretion—for my address and cell phone.

I never give out the home phone.

Never?

No. 'Cause it's like, with a cell, it's like, there's no reason. And what I do at home is that I, like, leave the cell phone in the basement when I don't want to hear it.

You don't turn it off?

No, I mean, there is no reason. Because if I turned it off, I'll forget that it's off the next day. 'Cause I don't look at it, I just pick it up.

So anyway, I leave it in the basement so it doesn't wake me up or anything.

But if you give them your home phone—You assume if people call you at midnight on the home phone, there's some emergency.

So it's okay for people to call you at midnight on the cell phone because you aren't going to hear it anyway?

That's right.

Do people know that?

Or if I happen to be working late, you know, then I'll pick it up. But—They know it if they need to know it.

The key here is that it's the participant who gets to decide whether or not they do.

Efforts to allocate different people to different communication channels often focus on the issue of how directly we want someone to be able contact us. Three factors seem to underlie decisions about this issue of directness and, therefore, the precise ways that participants want others to reach them: (1) who the sender is; (2) what the precise reason is she or he might be trying to reach them; and (3) how attentive the participant is in monitoring each of her or his communication channels. (This last is important because this is what creates a fairly predictable level of availability for any given channel.)

In fact, these three factors lead to a hierarchy of preferred and less preferred channels for each sender and/or situation. Recipients may engage multiple phone lines and multiple email accounts as well as an entire smorgasbord of access points that cross multiple technologies to increase their demand-organizing and responding options. When used strategically, each channel can be used not only to better schedule the things and people one attends to, but also to provide a recipient with a sense of how important a certain demand might be, simply because of the way it arrives. Marshall McLuhan's (1965) insight that "the medium is the message" takes a very practical turn in just this way. By reserving a particular channel for communication with the most important people in their lives, for instance, or for the most important situations, participants know that when that channel is activated, they should respond to it as quickly as possible.

Once a demand has been sent down a particular channel, it may be further filtered before reaching the intended recipient. A child, spouse, or assistant may answer the phone first. A spam filter may assess email first. A cell phone may identify the sender's number and either display the sender's name or initiate a certain ring tone to signal a demand's arrival. A domestic phone service may block the call if it does not recognize the sender as a "white-listed," preapproved caller.

Figure 3.1 illustrates how such a system might work for a typical person in this study across the five channels commonly reported (i.e., cell phone, home phone, work phone, plus work and personal email accounts).

As shown in the diagram, a demand may be filtered one or more times after entering a specific channel. Afterward, if the demand continues on its journey toward the recipient, it may then be relegated into any number of holding places (e.g., a special spot on a desk for message slips, a chair, a refrigerator, an email inbox, a spam box, an answering machine or voice-mail, the mind of a husband who monitors a couple's joint email account, or the mind of a teenager who listened to the answering machine messages upon arriving home from school. Here, the accumulated demands await the designated recipient's attention.

In general, the more discretion a channel provides in terms of concealing one's availability, the more participants like it. The ability to conceal availability is a function both of the actual technological features of the channel and the kinds of use patterns an individual establishes around it.

As a result of both, users can exert a certain amount of control in any given situation as they seek to respond quickly to certain demands, but to put off others as they wish, too.

Toward this end, quite a few participants choose to reserve a specific communication channel for the most important people in their lives—or for the most important matters. Because of the highly personal and direct nature of the channel, cell phones were the commonly preferred channels reserved for this purpose. Cell phones and their features are inherently designed for an individual, able to be taken almost anywhere, and commonly carried close to the body. It allows the most ubiquitous, direct access to an individual. It is also a fashion, coming-of-age, and status accessory that has set new standards for conspicuous consumption (Veblen 1994). In the States, it is thought of as a highly personal object for all these reasons.[30]

Many participants had a difficult time envisioning, much less recalling, anyone but themselves ever answering their cell phones. Its association with only one specific individual means that people are rarely tempted to pick up, much less use, another's mobile phone—unless directly asked to do so.

Allocating different people to different channels is thus an important way in which participants try to counteract the boundary-blurring potential of otherwise ubiquitous accessibility. It helps reimpose spatial and temporal boundaries around what and whom they pay attention to—and when—in an attempt to regulate the selves, social networks, and practical tasks that might otherwise be activated at any and all times. Of course, not everyone may appreciate participants' attempts to place certain boundaries around them, particularly if it means rejecting their own preferred ways of getting in touch.

Channel Conflict

When people fail to align over their preferred channels of communication, conflict may well ensue. From minor annoyances to unmasked power struggles, contests over who does and does not have control, who will and will not bend in a relationship may be seen here. Individuals describe employing a variety of tactics, occasionally subversive, to condition others into using their preferred channels of communication—in the ways they prefer them used.

Refusing to use a particular channel, for instance, is a very effective way to get others to use yours. Here is an office assistant:

> Like I said, I'm a phone person. Even when they send me emails through the system, I don't normally send them an email back. I will pick up the phone and call them back and discuss it with them. ... Now, no one emails me. *(Laughter)*

And here is a mother, another email resister:

> **Okay, How many times a day or a week would you log onto your email account?**
>
> Maybe, if I remember, every week or so.
>
> **Uh huh.**
>
> I really don't log on as much as I should, and I have been yelled at by my son. Yes. *(Laughing.)* It is a good way of keeping in touch with people, it's just that it's so easy—to not turn it on.
>
> **Uh huh. I understand that. Are there ever any times when you don't return email messages that you receive?**
>
> Uh huh. *(Laughing.)* Oh, yes. Yes.
>
> **Is there a pattern to that? Or—?**
>
> My son says that there is a pattern. It is so easy to just put it off. I will do it later. I will do it later.
>
> **Uh huh.**
>
> Or if—He got desperate and he just finally picked up the telephone and said, "You didn't pick up your mail." And I hadn't. And that was it. I just hadn't. I just hadn't even opened it. So I know that I've got to do better.

Or not. What she does in the future may mostly depend on just how much her son pushes her on this—or just how badly she needs to reach him should he start refusing to answer the phone.

In fact, refusing to use or develop reliable accessibility via a certain channel may be a great way to make others' demands more manageable. The office assistant mentioned previously, for instance, not only has to pay less attention to email than others do in her institution, but as a result of this, gets asked to do far less than others do, too.

Corrupted Channels

Occasionally communication channels become corrupted and no longer serve the function they once did. As we have seen, answering the home telephone before tele-marketers became rampant was a far different experience from what it is now. Because of these solicitors, this communication channel has certainly lost at least a measure of directness and privateness for everyone in this study. It has been corrupted. The home phone's function—as well as people's interactions with it—has changed accordingly.

Any time an "intruder" appears within a specific channel, it can remind us of our preferences for that channel. Here, one participant recalls a story about her relationship to instant messaging and the bind she finds herself in when a family member begins popping up there.

> **Have you ever monitored someone else's computer use? And this also includes friendly monitoring, like instant messaging, you know, when you're looking online to see if somebody else is online.**
> No. I so don't care about that. In fact, I find that really irritating—that some-one can find out if I'm online and send me an instant message that I have to respond to. Actually, the only person who does that to me is (my niece) and I can't very well not reply to my niece, so—And I'm working, and then I get these messages from her and that's really irritating. Luckily, she's the only person that does that to me.

If the channel becomes corrupted to the point where the annoyances and intru-sions associated with it outweigh the benefits of using that channel, it can foster such a negative reaction that the channel may be abandoned altogether. People who have no landline at home because of telemarketers and use only their cell phones fall into this category.

Stories of mistakes made and lessons learned in the "struggle to achieve a balance between privacy and participation" (Westin, 1967, 11) abound for these participants. Rather than simply giving up, though, for the most part—and in spite of a palpable sense of frustration—most of them seem to be holding their own. They do so in the ways that people usually do: by taking the resources at hand and putting them to work as best they are able in order to address the problem. In the process, they reveal much about the current problems, solutions, and tradeoffs of managing social accessibility in an information and communication technology-saturated world.

Conclusion: Technology and Privacy

Technology[36] provides an entry point from which we can uncover what is socially prob-lematic. It gives us a place from which we can see what a social group takes for granted as well as its more contested terrains. Technology is itself a kind of hyperlink that can take us into the assumptions and practices surrounding it.

This characteristic of technology persists at least in part because of the way it functions as a prosthetic.[37] Technology allows us to extend our senses, our reach, our influence, and other aspects of our selves. It does this by extending our physical capabilities (including our presence) over time, space, and social networks of relationships.

These two interrelated characteristics of technology mean that as new and more powerful technologies are made—and older ones are abandoned—we bring them with us into experiential realms and daily routines where they did not previously appear. As a result, cultural concepts and expectations become juxtaposed and linked in new ways with specific times, spaces, people, objects, and activities. In the process, we not only find new challenges to our previous understandings and experiences of the world, but we also discover that much of what we had taken for granted was and is largely socially constructed, however reified it may have become.[38]

This chapter has focused on the ways this process is played out regarding privacy and the challenge of managing social accessibility. Changing technologies, expectations, and the habits that incorporate them mean that the need to attend to the problem of social accessibility is highly likely to persist in the future. The need to personally attend to this aspect of achieving privacy certainly will not go away.

This is not only because of practical problems like where to focus one's attention given all the people who want it and who can now knock on one's electronic as well as physical doors. Gergen (1991) believes that the very concept of self is becoming defined more thoroughly by relationships because of these changes, too—relationships that are ever increasing in number as well as importance.

> In this era the self is redefined as no longer an essence in itself, but relational. In the postmodern world, selves may become the manifestations of relationships, thus placing relationships in the central position occupied by the individual self for the last several hundred years of Western history.

Our ability to manage those relationships via the management of our accessibility may take on even greater importance in the future, accordingly—whether it's in response to the phone ringing, an email or text message arriving, some new and currently unimaginable technologies, or the person sitting next to us who is competing with all of these.

FYI: TMI: Toward a Holistic Social Theory of Information Overload

By Anthony Lincoln

Introduction

Both the popular and academic presses warn of the growing problems created by an ever–increasing flow of information. The *Economist* (2010) warns us of "monstrous amounts of data" in a new special report. A recent article in *IEEE Spectrum* refers to our "infoglut" as "the disease of the new millennium"[1]. CNN, in turn, pleads, "How can we cope with information overload?" (Mollman, 2010). It would seem that information overload—or the cognitive overload to which information superabundance contributes—continues to track the growth curve of information itself, distracting attention, hampering decision–making, and lowering productivity in and out of the workplace. Interest, often business–related, in isolating the causes and combating this "ill" in organizations has spawned copious amounts of research, working groups, and indeed entire industries (decision support, knowledge management, social networking tools, etc.) committed to the fight.

Research has been extensive and cross–disciplinary, producing a multitude of suggested causes and posed solutions. I argue that many of the conclusions arrived at by existing research, while laudable in their inventiveness and/or practicality, miss the mark by viewing information overload as a problem that can be understood (or even solved) by purely rational means. Such a perspective lacks a critical understanding in human information usage: much in the same way that economic models dependent on rationality for their explanations or projections fail (often spectacularly, as recent history attests), models that rely too heavily upon the same rational behavior, and not heavily enough upon the interplay of actual social dynamics—power, reputation, norms, and others—in their attempts to explain, project, or address information overload prove bankrupt as well. Furthermore, even research that displays greater awareness of the social context in which overload exists often reveals a similar rationality in its conceptualization. That is, often the same "social" approaches that offer potential advantages (in mitigating information overload) over their "non–social" counterparts paradoxically raise new problems, requiring a reappraisal of overload that takes social issues into account holistically.

"Information," Scope, and Volume

Mason, *et al.* (1995) describe an "epistemic hierarchy" in which information is produced. First the mind draws distinctions and creates data out of chaos. Such data, categorized, organized, and rationalized based on the processing mind's perspective, undergoes a transformation into information. Subsequent scrutiny to authenticate or verify it results in knowledge. Beyond that, it becomes broadly codified and integrated both with knowledge from other disciplines and with a culture or society. The resultant byproduct, wisdom, then, is both figuratively and literally the farthest thing from data, involving "forgetting as much as remembering and is made up of insights and understandings as to what is true, right, and lasting"[2]. This paper deals primarily with data and information, the two of which are often conflated[3]. It also examines information overload chiefly as it occurs within the organizational sphere, in part because of the abundance of research pertaining to this context, but also because of the greater standardization of information use and thus the increased potential to draw general conclusions from it. However, a judicious amount of non–business examples present themselves in a small amount of relevant cases. Lastly, this paper does examine multiple overload "categories," in part because they contribute insight germane to the overarching conversation, and in part due to the ambiguities surrounding the concepts and terminology themselves.

Definitions, Concepts, and Contexts

Much like art or obscenity, the concept of information overload is difficult to define, but we "can all recognize the condition … when we see it"[4]. Less subjective definitions exist; one of the clearer ones notes that information overload conveys "the simple notion of *receiving too much information*"[5]. Scholarly terminology, runs the gamut: "cognitive overload (Vollman, 1991), sensory overload (Libowski, 1975), communication overload (Meier, 1963), knowledge overload (Hunt and Newman, 1997), and information fatigue syndrome (Wurman, 2001)"[6]. Edmunds and Morris (2000) add "analysis paralysis" to the list[7]. On top of this, Wurman argues that such overload leads, paradoxically, to information anxiety, the reaction to the gap between all the information we understand and what we think we ought to understand, the "black hole between data and knowledge"[8]. Others suggest "interaction" or "transaction overload" (Mathiassen and

Sørensen, 2002). More recently, researchers advocate the broader rubric of "technology overload," under which would fall "system feature overload," communication overload, and information overload (Karr–Wisniewski and Lu, 2010).

Whether examined broadly or constrained within a single rubric such as business or management, the context in which information overload occurs varies widely. "Information overload is frequently referred to in the literature of a range of disciplines such as medicine, business studies, and the social sciences as well as in computing and information science"[9]. Within an organization, overload has been examined among decision–makers as well as individual contributors, as well as within specific business areas (accounting, management of information systems (MIS), organizational science, and marketing) (Eppler and Mengis, 2004). Edmunds and Morris note that Butcher (1998) details three dimensions of management research into information overload: personal information overload and implications on problem–solving and decision–making; organizational information overload in which a surfeit of "documents" creates a sclerotic effect on productivity; and, customer information overload's effect on spending habits.

Conceptual approaches that attempt to quantify information overload show even greater disparity. Eppler and Mengis (2004) outline four general categories: characteristics of the information itself (both qualitative and quantitative); limitations in the individual's ability to process information (compared to the amount of information received); organizational issues such as formal or informal processes; and, information technology characteristics that govern how information is generated, transmitted, and received. In addition, subjective approaches examine individuals' emotional responses (anxiety, confusion, low motivation) to information overload as a qualitative measure of its effects.

Individual cognitive capacity is often presented as a central cause of information overload. Miller's seminal work illuminates human limitations both in "bandwidth" and in numeric processing: a human can process about seven "chunks" of data at a time[10], and tends to *subitize* items in groups of fewer than seven but estimate items in groups of greater than seven (Miller, 1956). Other cognitive bounds may benefit from a computer science conceptualization. Multitasking, frequently cited as a contributing factor in information overload, may be thought of in terms of a central processing unit's ability to handle different tasks simultaneously[11]. Further, interruptions and distractions, also examined in overload analyses, can be considered analogous to the context switch that a computer must undergo every time it sets aside one task and returns to another. Each of these limitations brings a quantifiable cost to bear on the individual's information processing capacity.

Characteristics of the information itself may produce or exacerbate overload. Issues of volume or quantity, where supply exceeds processing capacity, pose problems (Eppler and Mengis, 2004). Data rate plays into this as well, as do signal–to–noise dynamics as observed by Klapp (Edmunds and Morris, 2000). Schick's and Lawrence's temporal approach causally links overload to supply (or, more specifically, the amount of information processing required) and the amount of time an individual has at his disposal to perform such processing (Schick, *et al.*, 1990). Information (or even data) organization factors may affect information overload as well (Zaki and Hoffman, 1988). Qualitative traits, (Eppler and Mengis, 2004), such as uncertainty, diversity, ambiguity, novelty, complexity, intensity, quality or value, can all result in overload. Usage patterns or preferences such as multitasking or polychronicity may also influence factors affecting overload (Hecht and Allen, 2005).

Organizational factors may present themselves in information overload. If organizations can be considered information processing systems (O'Reilly III, 1980), they become subject to some of the same cognitive "computing" limitations outlined earlier with regard to individual causes of information overload. Yet in paradoxical analogy to a high-performing organization exhibiting strength greater than the sum of its parts, organizations may actually display magnified susceptibility to information overload. In addition to individual causes of overload, the need among managers to share, verify, or preemptively store information contributes to overload (Edmunds and Morris, 2000). More generally, organizational conditions such as increased collaboration, information centralization (or, conversely, disintermediation) may play a role as well (Eppler and Mengis, 2004).

Lastly, few would argue against technology's impact on information overload, although disparity persists over whether that impact has been, on the whole, positive or negative. It is perhaps the only cause or contributor to information overload also used, paradoxically, as a tool to mitigate the problem it helped create (Schultz and Vandenbosch, 1998). A large body of research on systems such as e-mail exists; as early as nearly a quarter century ago, researchers hypothesized that e-mail not only accelerates the exchange of information, but also leads to the exchange of new information too (Sproull and Kiesler, 1986). Many other e-mail studies demonstrate quantitative and qualitative impacts (Edmunds and Morris, 2000). Extensive analysis has also been performed on so-called "push" technologies, Internet/intranet/extranet deployments, high-storage capacity, lower duplication costs, and speed of access (Eppler and Mengis, 2004).

Past as Prologue: History, Symptoms, and Effects

The symptoms, if not the terminology, of information overload are hardly novel, dating back at least as far as the late nineteenth century, when indications of the burgeoning problem began to appear. In a recent letter to the editor of the *Chronicle of Higher Education*, the dean of an information school remarked that

> "[p]roblems associated with information overflow or overload have been of concern for a long time. In 1881, Dr. George Beard wrote American nervousness: Its causes and consequences, a supplement to nervous exhaustion. Beard believed that a chief cause of nervous exhaustion was the proliferation of reading material brought about by the invention of the high-speed

printing press in the nineteenth century. With the increased output of the periodical press—newspapers and magazines—there was suddenly too much to read in too little time. At its most benign, the nervousness could result in headaches or dyspepsia; but Beard warned that at its most acute, it could lead to insanity" (Cloonan, 2010).

Likewise, Edmunds and Morris (2000) point to the advertisement of a desk designed specifically for filing documents in the 1880s, and an account of the drastic growth in case law over the same period as examples. Without making overly light of the problem's seriousness at its humble origins, more pressing issues in the Industrial Age likely distracted attention from information overload, and the relatively high cost (and thus slow pace) of information technology innovation[12] arguably served to dampen the problem's growth. The proportion of any labor force employed specifically as information workers would have been miniscule—Shenk estimates four percent in 1850 (Edmunds and Morris, 2000), compared with 95 percent in 2000 (Mason, *et al.*, 1995)[13]. Technological advances including a steady doubling of processing power and the advent of the Internet have powered dramatic increases in the amount of information accessible and a counterintuitive pressure, particularly in business realms, to obtain more information (2000). The resulting abundance of—and desire for more (and/or higher quality)—information has come to be perceived in some circles, paradoxically, as the source of as much productivity loss as gain.

Categorized, the effects of information overload evoke a sort of overload of their own. Edmunds and Morris (2000) counterpose the abundance of information and the dearth of *useful* information: the promise of plentiful information, dimmed by the difficulty—and hence the cost in time and effort—of culling through it all in order to discover the items one needs. Overload can manifest itself in a number of ways: information retrieval limitations (*e.g.*, deteriorating search strategies, difficulty in identifying relevant information, problems reaching target audience), non–standard information processing and organization (*e.g.*, inconsistent and non–discrete categorization, insufficient analysis, misinterpretation), lower effectiveness in decision–making (*e.g.*, decreased quality or accuracy, reduced efficiency), or individual discomfort (*e.g.*, increased stress, increased acceptance of error, decrease in learning) (Eppler and Mengis, 2004)

A variety of countermeasures to information overload's effects exist, some prescribed, some observed, and some merely posited. Personal factors (*e.g.*, improved time management, augmented information literacy, improved personal information management), information characteristics (*e.g.*, increased quality, organization, visualization, or interfaces), task/process parameters (*e.g.*, uniform procedures, information handling strategies), organizational design (*e.g.*, coordination, hiring/scheduling decisions), and information technology—are all named as mitigating factors (Eppler and Mengis, 2004)[14]. Similarly, data delivered in summarized form (as opposed to raw) increased the quality of subsequent decisions (Chervany and Dickson, 1974), and arguably reduced overload (although there were unintended negative consequences as well). The need to keep current, and the requisite effort to do so, might be mitigated with systematic reviews of information (Edmunds and Morris, 2000). Some advocate technological approaches, which range from the practical—such as improving information retrieval techniques for greater precision and recall (Montebello, 1998), enhanced filters, or better mixes of "pushed" versus "pulled" information (Holtz, 2008)—to the fanciful, including intelligent agents (Mathiassen and Sørensen, 2002), Semantic Web designs (Breslin, et al., 2009), and more. Cognitive traits can work against overload as well. Higher amounts of polychronicity, when neither excessive nor deficient, can improve job fit and well–being (Hecht and Allen, 2005). Heavy media multitaskers show lower performance in task–switching than light media multitaskers (Ophir, et al., 2009). Hybrid solutions exist as well: in a study on a groupware implementation, the lack of increase in information overload is attributed to the technology itself and the human propensity toward selectivity as a filtering mechanism (Schultz and Vandenbosch, 1998).

Paradox Lost, Paradox Regained

In spite of decades of research yielding many strategies for mitigating information overload, the problem still exists to an extent that suggests intractability. Brynjolfsson's identification of the Productivity Paradox—the existence of an orders–of–magnitude increase in IT–delivered computing power with no consequent increase in business productivity (1993)—was explained and at least partially refuted in later years by, among others, Brynjolfsson himself (Karr–Wisniewski and Lu, 2010). Yet we find ourselves in a similar quandary now. The crush of ready information produces productivity losses and gains, both real and perceived, and many of the factors cited above can contribute to either effect. How have we come so far and yet achieved so little insight?

One might argue that the rate of technological change has simply outpaced human cognitive ability, that new tools and strategies to stem the flow of information do not scale to match the increase in volume. Some have postulated that new technologies (specifically, social media) have accretively added to the mess or even created an entirely new type of overload: "social information overload" (Passant, et al., 2009), and look to technological improvements for answers. However, as technology becomes increasingly tailored to better facilitate social paradigms, we cannot hope for insight into this paradox without examining the problem in a social context. A deterministic, rational approach, where people behave predictably according to preordained rationales, no longer works.

Homo Informaticus: Rationality and the "Information Person"

Fundamental modern economic theory, as conceptualized by John Stuart Mill and crystallized in Adam Smith's *Wealth of nations*, depends upon a conceit of agents who act in predictable ways. Economists' "basic unit of study"[15], then, became an archetype of a human individual embodying specific traits brought to bear upon the economy in question. Rudiments of most modern Western theory

> "were associated with the concept of economic man [16], the cause and consequence of economic activity. During the earliest periods economic man was 'a relatively low–level abstraction thought to be descriptive of human nature. This description stressed self–interestedness, the securing of pleasure and the avoidance of pain, and rational calculation based on excellent knowledge of market conditions'[17]"[18].

Simon (1955) highlights the idealized characteristics of economic man and the rationality whereby he acts: "This man is assumed to have knowledge of the relevant aspects of his environment which, if not absolutely complete, is at least impressively clear and voluminous. He is assumed also to have a well–organized and stable system of preferences, and a skill in computation that enables him to calculate, for the alternative courses of action that are available to him, which of these will permit him to reach the highest attainable point on his preference scale"[19]. Despite the instructive benefits of a simplified, rationalized research subject, conclusions drawn from it rely upon a significant set of implicit assumptions: a fixed *modus operandi* that reliably places the interests of one's self before those of others, a focus (per Mill) purely on pleasure as a "good", and an intellectual calculus that not only possesses keen market insight, but also makes efficacious use of such insight.

Such assumptions, as recent global economic woes aver, can prove problematic. An individual does not always behave "rationally," in any sense of the word, and moreover, even the possession of complete information does not guarantee that an individual will either inform himself completely or act optimally—much less predictably—given such information. Capurro (2005) observes that "there is no possibility for us to fill the gap between information and knowledge and, consequently, between trust and anxiety. There is no mood–free rational economy. Even more, moods are not the opposite to rationality but rationality itself is already in a mood of a knower who trusts (or not) sense data and his/her (imperfect) predicting capacity". Economic rationality (the term), then, is as much a construct as economic man; human rationality in itself is as dependent on mood as any other behavioral trait.

Cognizant of such failings, and dubious of the model's suitability as a keystone for subsequent theorization, Simon proposed a behavior model of rational choice to "replace the global rationality of economic man with a kind of rational behavior that is compatible with the access to information and the computational capacities that are actually possessed by organisms, including man, in the kinds of environments in which such organisms exist"[20]. Drawing in part from psychological theory of rational behavior, Simon constructed definitions of "approximate" rationality that aimed to create a more realistic actor within the model, a "choosing organism of limited knowledge and ability"[21]. This new model would take into account, among other things, variation in the information–gathering process, acknowledging both cost and the related extent of the process (given a non–zero cost). Economic man, he surmised, does not have to act rationally to the point of (absolute) optimization—he only has to act rational *enough*: "Under favorable circumstances, [a simple pay–off] procedure may require the individual to gather only a small amount of information—an insignificant part of the whole mapping ... If the search for an a having the desirable properties is successful, he is certain that he cannot better his choice by securing additional information"[22]. By injecting a step of ascertaining how far the information–gathering task must extend, Simon rendered the rational actor's decision–making process variable rather than fixed, streamlining the process under favorable conditions but complicating it otherwise.

Simon's subsequent work yielded the concept of "bounded rationality," the sense that our rationality is always influenced by other factors. "... [T]he execution of our rational capacities makes use of our resources, and our temporal, computational, and motivational resources, and whatever else we need for deciding, are always limited"[23] "Satisficing," taking just as much information as needed, has been described as both a common adjustment in overload situations and an exemplar of bounded rationality (Bawden and Robinson, 2008).

Presciently, Simon's Behavioral Model also envisions some of the socio–technical challenges still prevalent in the issues surrounding information overload. He notes that his attention to the information–gathering and –processing phases of decision–making may

> "suggest approaches to rational choice in areas that appear to be far beyond the capacities of existing or prospective computing equipment. The comparison of the I.Q. of a computer with that of a human being is very difficult. If one were to factor the scores made by each on a comprehensive intelligence test, one would undoubtedly find that in those factors on which the one scored as a genius the other would appear a moron—and conversely. A survey of possible definitions of rationality might suggest directions for the design and use of computing equipment with reasonably good scores on some of the factors of intelligence in which present computers are moronic."[24]

Aware of the strengths and weaknesses in both human and non–human computational abilities, Simon notes that each party's set of attributes complement those of the other's, raising important implications for design, not to mention use—anticipating to some small degree Ackerman's (2000) socio–technical gap.

Roberts' apt search for an analogous "information man" reflects just such a gap. Information man, he asserts, represents a basic unit of study for information users—essentially actors within an economy of information. Similar to economic man, information man displays the same habits and predilections (engaging in "rational" acts based on complete knowledge of information sources in seeking out "optimal" information, dwelling in an infocentric world shaped only by information use and unaffected by outside factors, and acting only within a conspicuously artificial environment such as a formal information system of a single organization). Information scientists, therefore, have conceptualized only the most primitive of information man, based on the same set of assumptions manifested in early economic man, and by theorizing on such a construct, have achieved only the same unenlightening results as economists basing theories on rudimentary concepts of rationality. "The frustratingly dead-end character of user studies based upon simplistic behavioural assumptions, and of quantitative work unillumined by systematically sought explanation, has led to developments which broadly parallel those observed in economics, although over a much shorter time span"[25]. Rigid exclusion of qualitative studies has prevented the capture of specific information about the user and the organization within which she interacts, preventing a fuller understanding of information man.

Attempts at a modern rationality as it applies to information overload proliferate, and many of them incorporate social dynamics. Some question basic assumptions about rationality itself in the design of information systems, arguing that the design process benefits from an increased awareness—and conscious manipulation—of symbolism as it pertains to an organization's socio-technical interactions (Hirschheim and Newman, 1991). Kumar, *et al.* (1998) expand on Kling's first and second rationalities (the first, an econo- and technocentric rationality where humans and systems work in harmony toward the economic interest of the organization; the second, a more bounded rationality that allows for investigation of human and social phenomena, in which power and politics play a role and on occasion, work against the interests of the greater organization) to construct a third way that takes into account the advances of the second rationality, but with an increased emphasis on cooperation and trust, which, in a real-world setting, emerged as the predominant values driving socioeconomic behavior (Kumar, *et al.*, 1998). However, although analyzing information economies through such a framework offers significant insight, it raises the same concerns as those associated with previous rationalities—even bounded rationality still projects the biases of the framework onto the object of analysis. Furthermore, such a framework displays worrisome similarities to the first rationality (in which humans and systems work harmoniously), except that now, instead basing itself on perfect cooperation, the model accepts a slightly less idealized version. By contrast, so-called "local rationality" examines a manager's propensity to apply rationality to pieces of information that may not be the ones most salient to a decision's broader context. Thus, the very presence of information may have dysfunctional consequences even if decision-makers do not process it incorrectly, suggesting that information's effect on decision-making and its effect on performance are mutually exclusive. On a larger scale, it offers insight as to why information-"naïve" organizations can outperform those with comparatively sophisticated information systems (Glazer, et al., 1992), and it underscores the vexing inconsistencies inherent in rationality-based analysis.

In a review of relevant literature, such perspectives have existed in the minority, marginalized (bounded?) by other rationalities, purely technical approaches, or cognitive analyses. Most analysis dependent upon social dynamics tends to display a form of "rationality" by focusing on a single social attribute or set of attributes, resulting in arguments underpinned by the implicit assumption that the particular dynamic studied exists in a vacuum, unaffected by other social factors (or, for that matter, technological or cognitive ones). Yet there exists a large enough body of literature to support a more holistic social theorization toward information overload.

Toward a Holistic Social Theory of Information Overload

As Brown and Duguid (2000) established a decade ago, technology design and use stand to gain when examined within their encapsulating social context. But as our information technology platform grows in sophistication, its closer approximations of social interaction tend to heighten rather than resolve many social issues. Existing research—even work not directly focused on information overload—shows that issues of cooperation, motivation, social networks, power/politics, reputation, knowledge sharing, notifications, and norms consistently reveal themselves in contexts of information use that affects overload. These issues often work in concert or at odds with one another, facilitating, magnifying, or counteracting each other's effects upon information overload.

In organizations, cooperation varies in situations of abundance versus scarcity, and yet there is no scholarly consensus on the correlation between scarcity and cooperation. Applying a social dilemma perspective to the question, Aquino and Reed (1998) suggest that scarcity in an organizational setting can create the perception of a divergence of interest between an individual and the group, which can result in competition and conflict that negatively impact the organization's operations. Such effects are moderated by two factors: ability of the members to communicate, and the distribution of access to the shared resources (Aquino and Reed, 1998). But in an information context, do such dynamics transfer reliably? Information is, in many cases, infinitely replicable, hence non-rivalrous, and in fact a potential catalyst of information generation. Should information overload, then, lead to greater amounts of cooperation? No; in fact, just as with physical resources, the opposite is likely true. A study done on knowledge sharing among managers in the People's Republic of China revealed that individual factors, more than other variables, had significant impacts: greed decreased sharing, and self-efficacy increased it (Lu, *et al.*, 2006).

Knowledge sharing, itself a social act—and one that can be fraught with ritualistic undertones (Traweek, 1988), MacKenzie with Spinardi (1996) brings its social characteristics to bear on information use and, by extension, overload. Lu, *et al.* (2006) also found that organizational support increased utilization of information and communication technologies (ICT) resulting in more knowledge sharing, which might influence overload either positively or negatively. Further, the distinction between types of knowledge can complicate things as well. High-quality information can have low value (Zhao, *et al.*, 2008) because of lack of relevance or other factors. A failure to capture or meaningfully render implicit (tacit) knowledge can result in an inability to learn from past experience (Zhao, *et al.*, 2008); unfortunately, organization-supported knowledge sharing through higher ICT utilization proves more effective in dissemination of explicit, rather than implicit, knowledge (Lu, *et al*, 2006). Tacit knowledge may also impair knowledge renewal (Rong and Grover, 2009). Increased awareness of the social aspects underpinning knowledge sharing, then, may aid in the act itself, although the sharing of implicit knowledge still poses greater challenges.

Social network characteristics and analysis play an important role in overload as well. Managers who spent more time gathering information were more likely to perceive a strategic issue in a context of uncertainty as a threat; this was mitigated by how diverse a body of information they found (Anderson and Nichols, 2007). It stands to reason, therefore, that the amount of reliance placed on one's social network and the composition of one's network (in terms of degree, tie strengths, etc.) would have powerful impacts upon the manager's ability to find information—and the level of diversity of the information retrieved—and thus her perception of the issue itself. Furthermore, the manager's level of information overload, both real and perceived, would exert significant weight in this equation as well: would a manager in an overload situation be more/equally/less likely to rely on her social network than upon other sources?

Anderson (2008) examines this question in an empirical study on managers' information gathering behaviors, positing that individual differences in motivation define a manager's willingness to maximize information–gathering benefits in their social networks. The study's results demonstrate that individual network characteristics do affect the information benefits one can derive, but that these effects are stronger for managers motivated to utilize them. The findings here imply that these two traits result in improved information gathering. Cognizant that better and/or more information is not a panacea, one can argue with some credibility that information overload, narrowly defined, might be mitigated by social network characteristics and the motivation to take advantage of them.

However, social qualities and motivation in and of themselves do not necessarily confer benefits. Robert and Dennis (2005) uncover an intriguing paradox about media containing variant levels of "social presence," the ability to convey the psychological impression that people are physically present. The use of so–called rich media high in social presence increases motivation but decreases the ability to process information, whereas lean media low in social presence decreases motivation but increases the ability to process information. Thus rich media (high in social presence) has the simultaneous, contradictory capacity to enhance and hamper performance. Such a paradox poses a fascinating question for investigators of information overload: if any medium of any level of social presence—from e-mail to face-to-face communication—raises the risk of information overload (either via decreased motivation or decreased processing capability), what can we infer about the pervasive nature of information overload itself?

Issues of power also present themselves prominently in the information overload equation, and the results are paradoxical as well. Perceived information overload intensity seems to have a stronger relation "to power distance than to the volume of written information or number of information transactions processed by an individual" (Kock, et al., 2009). This "information overload paradox" (Kock, *et al.*, 2009) highlights the lack of attention given to power relations in analyzing perceived overload. Similarly, knowledge renewal is strongly correlated with an IT individual's perception of her department's dynamism, career satisfaction (both intrinsic and extrinsic), tolerance of ambiguity, and level of "delegation" (knowledge sharing via organization structural means). Further, tacit knowledge, because of its extreme difficulty of transfer, may actually hamper knowledge renewal based on issues of authority, specifically, the perception that an individual does not truly "own" the knowledge and is instead merely a "user" of it (Rong and Grover, 2009). Elements of power relations run through motivation to renew knowledge: the size of one's social network, the relative dynamism of one's department, and the amount of tacit knowledge one possesses.

Reputation, too, can be seen through a lens of power. In an internal knowledge market flush with information, attention becomes the resource in contention (Hansen and Haas, 2001). The more selective an approach a knowledge supplier takes, filtering low quality information and providing only the highest quality items, the higher the supplier's reputation as a valued resource and the more attention it garners. For lower-quality resources, a vicious cycle ensues: the more information of lower utility it posts, the less time information–overloaded consumers can afford to spend culling through it all. Ultimately the consumers abandon the resource, putting their attention on a resource with higher–quality information. The low–quality resource's response is to publish more (lower–quality) information in an attempt to win back those overloaded consumers (Hansen and Haas, 2001). As it becomes more difficult to glean knowledge from the sheer amounts of information, therefore, the supplier with a reputation for the capability to minimize overload—by providing a lower volume of higher–quality information—exerts the most power.

Or does it? In a modern enterprise environment, such power assumes a multivalent quality. Zammuto, *et al.* (2007) postulate that simulation, an Enterprise 2.0 "affordance" that offers the capability to explore what–if scenarios, creates a kind of virtual data reality whose infinitely replicable nature may ease decision making or add to an accumulation of data. It "can favor or shape a variety of uses … from empowering action to information overload" (Zammuto, *et al.*, 2007). Speculative data models may result in higher–quality knowledge, but by their nature they require higher quantities as well, increasing information overload for all who seek to benefit from the knowledge, thus throwing off the old rationality of the competitive knowledge market. In such cases, power relations shift too: as business intelligence moves increasingly into the hands of information "consumers" (in the form of ad hoc simulative reporting), knowledge becomes decentralized and personalized, and the model becomes even more "pull"–oriented than before. Control of overload, therefore, reverts almost entirely back to the consumer in this case, although the greater technological power (in the form of virtually endless data "realities") increases its likelihood. Furthermore, issues of identity, authority and motivation could be expected to appear, as high–currency agents in the old knowledge economy adjust to the new equilibrium. Suppliers accustomed to attention (and its concomitant benefits) and consumers who enjoyed higher status/

productivity/self-perception based on their knowledge connections would inevitably undergo some type of transition, although the impact of such a change is not clear.

Similarly, questions abound when examining notifications and norms in this context. In today's overloaded work environment, multitasking during meetings via the use of smart phones and laptops has increased in prevalence. Individuals who perceive themselves as suffering from overload tend to "e-multitask" more frequently (Stephens and Davis, 2009). Equally or perhaps more significantly, individuals who observe others' e-multitasking behavior, and who deem that behavior acceptable, will increase their own meeting multitasking behaviors, and this reveals itself both at the individual and organizational levels (Stephens and Davis, 2009). This has important implications because unlike other overload-related behaviors, multitasking effectively may depend more on cognitive than social factors: working memory, fluid intelligence, and attention are predictors of multitasking performance, but polychronicity itself and extraversion are not (Konig, et al., 2005). This finding exemplifies one of the central paradoxes in the contemporary overloaded work environment: social pressures that provoke behavior perceived by the group or individual to mitigate problems, when in actuality such behavior either depends upon individual social/psychological/techno-logical characteristics for efficacy, or it exacerbates such problems outright.

Co-evolution of tool and user can contribute to such uncertainty as well. E-multi-tasking, discussed previously, can take the form of multi-communication, simultaneous overlapping conversations. This activity, now a commonplace interaction in business settings, is accompanied by its own set of norms and affected by perceptions about one's own effectiveness at it. As errors made during interactions may influence future interactions (Reinsch, et al., 2008), potential risks may be higher while, paradoxically, the act of multicommunication inherently divides—and thus reduces—attention paid to each conversation. Again, if attention is power in more rational contexts of overload, adaptations of both user and tool can turn such a notion on its head, changing the nature of overload itself.

To expand upon O'Reilly's (1980) notion of an IT organization as an information processing system, it is not just a tool for automating existing processes: it has become an enabler of changes to the organization itself that in turn lead to productivity gains (Zammuto, et al., 2007). The IT organization's evolution from the purely technical to the socio-technical has precipitated changes not only in the way information is gener-ated, shared, and gathered, but also in the way it is absorbed as well. In contrast to other learning theories such as behaviorism, cognitivism, or constructivism, *connectivism* reflects the exponential increase in knowledge itself and its rapidly shrinking "half-life", emphasizing the accelerating pace of knowledge acquisition and the increased role that social connections play (Siemens, 2004). Most individuals will work in multiple fields during their working life rather than a single one, necessitating at least one major retraining phase. Learning will be done more informally, via more ad hoc methods such as on-the-job (a concept decreasingly divorced from physical locale) training or by otherwise experiential means. Social networks, communities of practice, and the tech-nology required to facilitate such connections will increase in importance. The "pipe" will become central, rather than merely the content flowing within it (Siemens, 2004). Visible manifestations of our increasing interconnectedness in the workplace already exist, in the tools used as much as the way in which individuals and groups use them.

Evidence of connectivist principles (if not their formal use) is visible in socio-technical communities (STC), typically consisting of social relations between individuals or groups focused on a specific interest or problem, within a specific institution or organization (Jahnke, 2010). Moreover, STCs can decrease the complexity and information overload within an organization, possibly by enabling individuals to obtain only the information they need at a given time (Jahnke, 2010). By leveraging one's connection to a community generating knowledge in a particular area, an information consumer may receive a "low flow" of high-quality information. However, it is entirely possible that as an STC expands, the flow of information will increase beyond optimal (or even acceptable) thresholds, requiring the development of new techniques or strategies.

If multi-communication and connectivism represent signs of co-evolutionary change, one could envision that behavioral and cognitive shifts promoted by co-evolution have begun to alter the types of overload experienced within the workplace. Research has identified "supertaskers," individuals who demonstrate markedly superior performance in dual-tasking, suffering none of the ordinary performance penalties involved in switching from one task to the next (Watson and Strayer, 2010)[26]. Supertasking ability has not been linked definitively to neurological or genetic factors (Watson and Strayer, 2010); might we attribute this to co-evolution as well, and if so, how do individual propensities toward overload change in such a paradigm?

Conclusions

This paper cannot endeavor to be a complete assessment of information overload, related types, contributing factors, or potential solutions. It is certainly not the first to examine issues of information (or even information overload) within a social context, nor to re-examine rationality in this area. However, this work exists in the recognition that new perspectives on existing things can be illuminating, and in the hope that perhaps it may serve in such a way.

A proliferation of complex social variables interconnects the various manifestations and effects of information overload. As with any other dynamic system, a single change can initiate dramatic and far-reaching reverberations. Variables like the ones examined in this paper, then, form their own "social network," with the same uniqueness and capacity for metamorphosis. To zero in on individual components can push others out of the frame.

Analysis of social dynamics as they relate to information overload has shown a similar inclination toward narrowness, examining in depth a specific paradigm to the reduction or exclusion of others. While it has produced some instructive results, this lack of inclusion has been done at the expense of a broader understanding of information overload in a holistic social context. Curiously, research done in this way finds itself culpable of some of the same criticisms of rationality–based models that it in itself may have leveled. This is not to impugn the value of previous research on the topic; rather, it is to suggest a new way of thinking when it comes to overload, one that acknowledges the restrictiveness of examining specific social dynamics and the necessity for crosscutting analyses that seek to understand with greater clarity how the interplay between such variables affect the whole. Paradoxically, this may require expanding the scope of relevant research in order to view emergent patterns in all their completeness—resulting, perhaps, in its own new form of information overload to confront as well.

Notes

1. Zeldes, 2009, p. 30.
2. Mason, *et al.*, 1995, p. 52.
3. Despite the primacy and ubiquity of the phrase "information overload", many of the works surveyed in this paper use the terms "data," "information," and in some cases "knowledge" loosely and on occasion interchangeably. (Although "wisdom," mercifully, is not.) I will attempt to normalize where possible and rationalize when necessary.
4. Edmunds and Morris, 2000, p. 18.
5. Eppler and Mengis, 2004, p. 325.
6. *Op.cit.*
7. They also credit "information fatigue syndrome" to Oppenheim (1997).
8. Wurman, 2001, p. 14.
9. Edmunds and Morris, 2000, p. 18.
10. Miller utilizes "chunks" rather than bits because of phoneme differences in words.
11. This is, of course, not strictly true: time–division multiplexing, one of the more common ways for a computer to multitask, allocates slices of CPU time to each job in small enough quantities (typically on the order of microseconds) as to be imperceptible to a human operator.
12. Exemplified perhaps by Babbage's difference engine, a mechanical computer just a bit ahead of its time.
13. A plausible (if somewhat melodramatic) modern conclusion is that we are all information workers now.
14. Compared to perceived information overload, perceived *underload* correlates with improved decision–making, despite lower satisfaction (O'Reilly III, 1980).
15. Roberts, 1982, p. 94.
16. Thought initially to be a pejorative for Mill's concept (Wikipedia, n.d.).
17. Gould and Kolb, 1964, p. 223.
18. Roberts, 1982, p. 94.
19. Simon, 1955, p. 99.
20. *Op.cit.*
21. Simon, 1955, p. 114.
22. Simon, 1955, pp. 106–107.

23. Spohn, 2002, p. 12. But see Bolton and Ockenfels (2009), who show, conversely, that investigating perfect reputation systems may result in valuable insights despite their simplicity.
24. Simon, 1955, p. 114.
25. Roberts, 1982, p. 102.
26. These individuals bring to mind generational distinctions in tool utilization, epitomized by ubiquitous thumb–texting youths and their parents who hunt and peck with an index finger.

www.ingramcontent.com/pod-product-compliance
Lightning Source LLC
Chambersburg PA
CBHW061413210326
41598CB00035B/6194